建筑与都市系列丛书 | 世界建筑
Architecture and Urbanism Series | World Architecture

文筑国际 编译
Edited by CA-GROUP

SuperNormal
Architecture in the Netherlands
2010-2020

荷兰：
超常规建筑（2010-2020）

中国建筑工业出版社

图书在版编目（CIP）数据

荷兰：超常规建筑：2010-2020 = SuperNormal Architecture in the Netherlands 2010-2020：汉英对照 / 文筑国际 CA-GROUP 编译 . -- 北京：中国建筑工业出版社，2021.1

（建筑与都市系列丛书 . 世界建筑）

ISBN 978-7-112-25724-9

Ⅰ.①荷… Ⅱ.①文… Ⅲ.①建筑艺术－介绍－荷兰－汉、英 Ⅳ.① TU-865.63

中国版本图书馆 CIP 数据核字 (2020) 第 247370 号

责任编辑：毕凤鸣　刘文昕
版式设计：文筑国际
责任校对：王 烨

建筑与都市系列丛书｜世界建筑
Architecture and Urbanism Series ｜ World Architecture

荷兰：超常规建筑（2010-2020）
SuperNormal Architecture in the Netherlands 2010-2020

文筑国际　编译
Edited by CA-GROUP

*

中国建筑工业出版社出版、发行（北京海淀三里河路 9 号）
各地新华书店、建筑书店经销
北京雅昌艺术印刷有限公司 制版、印刷

*

开本：787 毫米×1092 毫米　1/16　印张：17　字数：367 千字
2024 年 7 月第一版　　2024 年 7 月第一次印刷
定价：**148.00** 元
ISBN 978-7-112-25724-9
　　　　（36510）

版权所有　翻印必究

如有内容及印装质量问题，请联系本社读者服务中心退换
电话：（010）58337283　QQ：2885381756
（地址：北京海淀三里河路 9 号中国建筑工业出版社 604 室　邮政编码 100037）

a+u

建筑与都市系列丛书学术委员会
Academic Board Members of Architecture and Urbanism Series

委员会顾问 Advisors
郑时龄 ZHENG Shiling　崔 愷 CUI Kai　孙继伟 SUN Jiwei

委员会主任 Director of the Academic Board
李翔宁 LI Xiangning

委员会成员 Academic Board
曹嘉明 CAO Jiaming　　张永和 CHANG Yungho　　方　海 FANG Hai
韩林飞 HAN Linfei　　　刘克成 LIU Kecheng　　马岩松 MA Yansong
裴　钊 PEI Zhao　　　　阮　昕 RUAN Xing　　　王　飞 WANG Fei
王　澍 WANG Shu　　　　赵　扬 ZHAO Yang　　　朱　锫 ZHU Pei

*委员会成员按汉语拼音排序（左起）
Academic board members are ranked in pinyin order from left.

建筑与都市系列丛书
Architecture and Urbanism Series

总策划 Production
国际建筑联盟 IAM　文筑国际 CA-GROUP

出品人 Publisher
马卫东 MA Weidong

总策划人/总监制 Executive Producer
马卫东 MA Weidong

内容担当 Editor in charge
吴瑞香 WU Ruixiang

助理 Assistants
卢亭羽 LU Tingyu　杨 文 YANG Wen　杨紫薇 YANG Ziwei

翻译 Translators
英译中 Chinese Translation from English:
盛 洋 SHENG Yang (pp.12-19, 30-37, 84-93, 184-191, 246-253)
熊赫男 XIONG He'nan (pp.38-83, 94-133)
王霄晗 WANG Xiaohan (pp.134-175)
南方 NAN Fang (pp.198-213)
英译中 Chinese Translation from English:
杨紫薇 YANG Ziwei (p266)
日译中 Chinese Translation from Japanese:
吴瑞香 WU Ruixiang (p11, pp.220-245)　陈旭丹 CHEN Xudan (p268)

书籍设计 Book Design
文筑国际 CA-GROUP

中日邦交正常化50周年纪念项目
The 50th Anniversary of the Normalization of
China-Japan Diplomatic Relations

本系列丛书部分内容选自A+U第592号（2020年01月号）特辑
原版书名：
スーパーノーマル オランダの建築 2010～2020
SuperNormal: Architecture in the Netherlands 2010-2020
著作权归属A+U Publishing Co., Ltd. 2020

A+U Publishing Co., Ltd.
发行人／主编：吉田信之
客座主编：克丝汀·汉内马
副主编：横山圭　Sylvia Chen
编辑：Grace Hon　小野寺谅朔　佐藤绫子　Cameron Cortez
海外协助：侯 蕾

Part of this series is selected from the original a+u No. 592 (20:01), the original title is:
スーパーノーマル オランダの建築 2010～2020
SuperNormal: Architecture in the Netherlands 2010-2020
The copyright of this part is owned by A+U Publishing Co., Ltd. 2020

A+U Publishing Co., Ltd.
Publisher / Chief Editor: Nobuyuki Yoshida
Guest Editor: Kirsten Hannema
Senior Editor: Kei Yokoyama Sylvia Chen
Editorial Staff: Grace Hong Ryosaku Onodera Ayako Sato Cameron Cortez
Oversea Assistant: HOU Lei

封面图：莱克纳尔博物馆的北立面，建筑设计：哈佩尔– 科尼利斯– 韦尔霍芬建筑师事务所
封底图：劳克哈尔图书馆室内，建筑设计：城市建筑事务所

本书第264页至271页内容由安藤忠雄建筑研究所与震旦博物馆提供，在此表示特别感谢。

本系列丛书著作权归属文筑国际，未经允许不得转载。本书授权中国建筑工业出版社出版、发行。

Front cover: North facade of Museum De Lakenhal, designed by Happel Cornelisse Verhoeven. Photo by Karin Borghouts.
Back cover: Interior view of LocHal Library, designed by CIVIC Architects. Photo by Stijn Bollaert.

The contents from pages 264 to 271 of this book were provided by Tadao Ando Architect & Associates and Aurora Museum. We would like to express our special thanks here.

The copyright of this series is owned by CA-GROUP. No reproduction without permission. This book is authorized to be published and distributed by China Architecture & Building Press.

Preface:
The Global Spirit of Dutch Architects

MA Yansong

During the economic boom and rapid urbanization before 2008 Beijing Olympics, a group of Dutch architects came to China. They almost seemed like ambassadors of globalization, active in major cities all around the world with creative proposals. Their design combines rationality and craziness: rationality is inherited from modernism principles, as logic and reason are always the core of their design; craziness comes from the cultural genes of the Netherlands, which are equality, freedom and openness. In the mass media era, their designs are eye-catching with strong visual impressions and bold expressions.

In 2011, MAD was commissioned to make a conceptual masterplan proposal of 200,000 square meters urban mixed-used. We designed 11 mountain-like towers, covering the site with organic curves. The proposal aroused a lot of noises and did not get the planning permit. Two years later in the competition for Boijmans van Beuningen Museum expansion, MAD was shortlisted as the only Chinese architecture firm besides the other four Dutch firms. It was a really tough competition and MAD ranked the second. After a year of drama and lawsuits between the first ranked firm and the organizer, the winner kept the same.

That was the first time we encountered Dutch firms in such competitive scenarios, which deepened my understanding of the Netherlands. Comparatively, Amsterdam is more conservative and Rotterdam more radical. It makes me ask that why Rotterdam has so many globally famous architects? As modernism become a leading theory in the post-war era, the Dutch architects expand the legacy of modernism and create a highly organized theoretical framework. All the Dutch architects challenge and contribute to this framework, making it evolves continuously and organically. As a result, Dutch architecture grows from the same soil but blooms differently. They transform architecture into a universal language that expresses Dutch culture and spirit. With the Dutch architects' active participation in the modernization process all around the world, Dutch architecture almost becomes an equivalent to global architecture, pioneering in the global architectural practices.

Many years later, Droom en Daad foundation commissioned MAD to design FENIX Museum of Migration in Rotterdam. Originally built in 1923, Fenix was once the largest warehouse in the world. It is adjacent to the largest port in Europe, connecting Rotterdam with the world. Millions of Europeans boarded here to start a new life in America, and many Chinese immigrates first settled down here. Through this project, I observed and understood Rotterdam more in depth. The adventurous spirit has long been scripted in the local history. Curiosity and hope still drive the locals to move on and take new challenges. That is also how the Dutch architects make their presence unignorable in the global scene, having their heart beats resonate with the whole world. This is the spirit of Dutch architecture for me.

This is very inspiring to Chinese architects. Chinese architects tend to focus on local issues, and I think these issues can generate world-wide impacts. We need to confront global issues and practice all around the world with more courage, just as the Dutch architects.

序言:
荷兰建筑师的全球精神给中国建筑师的启示

马岩松

在经济蓬勃发展、城市高速开发的"前奥运"时代，一群荷兰明星建筑事务所涌入中国。彼时，荷兰建筑师仿佛是全球化的使者。他们活跃在中国和世界各大城市，提出各种奇思妙想，既冷静又疯狂。冷静来自于现代主义的延续、务实，从根本上对理性和逻辑的推崇；疯狂来自于荷兰所推崇的平等、自由和开放，人性的解放。媒体时代，他们的设计很容易成为焦点，夸张、自信、个性张扬。

2011年，MAD被委托在阿姆斯特丹做20万平方米城市综合体的规划。我们设计了如同延绵不断的山峰一般的11座塔楼，有机的曲线爬满整个地块。方案引起轩然大波。在一片激烈的讨论声中，规划报批不了了之。两年后，在鹿特丹博伊曼斯·范伯宁恩美术馆（Boijmans van Beuningen Museum）的加建竞赛中，MAD成为最终五家事务所中唯一的中国建筑师事务所，与四组荷兰建筑师事务所比赛。对手强劲，竞争激烈。方案评选我们排名第二，没想到剧情转折太快，跌宕起伏，排第一的事务所跟主办方打了一年官司才最终拿到设计权。

这是我们与荷兰建筑师最直接的一次交锋。经过这两次来往，加深了我对荷兰的了解。阿姆斯特丹相对保守、传统，鹿特丹反而更新潮、开放。为什么鹿特丹这么小的一座城市，培养出这么多世界著名建筑师？顺应战后现代主义的崛起成为"先进文化"，建筑师们搭建了严谨的理论体系，持续推演，相互补点，大胆实践。可能也没有哪个国家的建筑师有如此清晰的群体特征，一脉相承又各具特色。建筑成了他们通用的语言，成为荷兰文化和国家精神的外化表达。他们游走在世界各地参与着现代化的进程，把荷兰建筑同"国际化"画了等号，成为全球范围实践先锋的代名词。

在这之后很多年，Droom en Daad基金会委托MAD设计鹿特丹移民博物馆（FENIX Museum of Migration）。移民博物馆选址在Fenix，原建造于1923年，曾经是全世界最大的仓库，坐临欧洲最大港口，连接世界。千百万欧洲人曾于此登船前往美国，也有大批华人从这里开始侨居欧洲。通过这个项目，我更仔细观察和理解了这座城市，心中的疑问算是得到解答。对新世界的不断探索就写在他们的历史里，对未知的好奇和希望刻在灵魂中，驱动着他们不断前进。荷兰建筑师们愿意把自己的命运与世界联系在一起，热切地关注全球范围发生的事情。我想这可能就是荷兰建筑的精神。

这对中国建筑师充满启发。我们倾向于关心"自我"和"本土"等非常具象的问题，然而今天中国的本土问题足以产生世界范围的影响。在这样的语境下，中国建筑师是否应当去理解当下世界面临的困境和危机，像荷兰建筑师一样，勇敢实践，不断前行。

MA Yansong

Founder & Principal Partner, MAD Architects. Guest Professor in Beijing University of Civil Engineering and Architecture, Tsinghua University, University of South California.

马岩松

MAD建筑事务所创始人、合伙人。曾任北京建筑大学、清华大学、美国南加州大学客座教授。

SuperNormal Architecture in the Netherlands 2010–2020

Preface:
The Global Spirit of Dutch Architects   6
MA Yansong

Introduction:
SuperNormal   12
Kirsten Hannema

Essay:
Reshaping the Polder   30
Bob Witman

bureau SLA, ZakenMaker
Co-living Oosterwold   38

Space&Matter
De Ceuvel   48

Hilberinkbosch Architects
The Sixteen-Oak Barn   58

RAU architects, RO&AD architects
Tij Observatory   68

RAAAF, Atelier de Lyon
Deltawerk//   76

Essay:
Tabula Scripta   84
Jarrik Ouburg

Robert A.M. Stern Architects (RAMSA), Rijnboutt
Beurspassage – Oersoep   94

Superuse Studios
BlueCity Rotterdam   102

Neutelings Riedijk Architects
Deventer City Hall   110

CIVIC Architects
LocHal Library   122

Neutelings Riedijk Architects
Naturalis Biodiversity Center   134

KAAN Architects
Education Center Erasmus MC   144

Frits van Dongen Architects and Planners, Koschuch Architects
Musis Sacrum   156

Happel Cornelisse Verhoeven
Museum De Lakenhal   166

NL Architects, XVW Architecture
Kleiburg De Flat   176

Essay:
After the NAi: Grand Projects in the Netherlands   184
Sergio M. Figueiredo

NL Architects
Forum Groningen   192

Ector Hoogstad Architects
Utrecht Bicycle Parking   200

Mecanoo
Delft City Hall and Train Station   208

Koen van Velzen Architects
Public Transport Terminal Breda   220

Cruz y Ortiz Architects
The Rijksmuseum   232

An Interview with Floris Alkemade:
The Future of the Netherlands   246
Kirsten Hannema

Architects Profile   254

Spotlight:
Aurora Museum   264
Tadao Ando Architect & Associates

荷兰：超常规建筑（2010–2020）

序言：
荷兰建筑师的全球精神给中国建筑师的启示　7
马岩松

导读：
超常规　12
克丝汀·汉内马

论文：
重塑围垦地　30
鲍勃·惠特曼

布洛 SLA，扎克马克
奥斯特沃尔德共享住宅　38

空间 & 变体建筑事务所
德·苏维尔　48

希尔伯林克·博世建筑事务所
十六号橡树谷仓　58

RAU 建筑师事务所，RO&AD 建筑师事务所
Tij 鸟类观测站　68

RAAAF，里昂工作室
三角洲之作 //　76

论文：
Tabula Scripta　84
贾里克·奥伯格

罗伯特·A.M. 斯坦恩建筑事务所，里恩博特
展览之廊 - 奥索普　94

超级利用工作室
鹿特丹蓝色城市　102

诺特林斯·里迪克建筑事务所
代芬特尔市政厅　110

城市建筑事务所
劳克哈尔图书馆　122

诺特林斯·里迪克建筑事务所
荷兰国家生物多样性中心　134

KAAN 建筑师事务所
伊拉姆斯大学医学教育研究中心　144

弗里茨·范·东恩建筑规划事务所，科舒奇建筑事务所
圣珂兰音乐厅　156

哈佩尔 - 科尼利斯 - 韦尔霍芬建筑事务所
莱克纳尔博物馆　166

NL 建筑师事务所，XVW 建筑师事务所
克莱堡公寓　176

论文：
NAi 之后：荷兰的宏大项目　184
塞尔吉奥·M. 菲格雷多

NL 建筑师事务所
格罗宁根广场　192

埃克特·胡格斯塔德建筑事务所
乌得勒支自行车车库　200

麦肯诺建筑事务所
代尔夫特市政厅和火车站　208

科恩·范·维尔森建筑事务所
布雷达汽车站与火车站　220

克鲁兹·奥尔蒂斯建筑事务所
荷兰国家博物馆　232

采访弗洛里斯·阿尔克马德：
荷兰的未来　246
克丝汀·汉内马

建筑师简介　254

特别收录：
震旦博物馆　264
安藤忠雄建筑研究所

Editor's Word

编者的话

In this book, together with guest editor Kirsten Hannema, Chief Editor of the nai010 Yearbook Architecture in the Netherlands, we will take a look at the recent 10 years of Dutch architecture discourse. Previously in our a+u 12:01 issue (Feature: Architecture in the Netherlands 2000–2011) highlighting the country's architecture scene right after the global financial crisis in 2008, the construction industry faced a slowdown and what we saw then as "typically" Dutch, has gradually faded. Looking away from the "SuperDutch" stigma, we now see projects combining broader social issues and since, take on a new form of cultural energy. Yet, compared to their previous superstar generation, these architects and their projects often go unnoticed. Therefore, in this book, 19 projects are selected and placed into 3 themes to introduce to you a new attitude on today's architecture in the Netherlands – the "SuperNormal".

The 3 themes: "Reshaping the Polder" looks into pilot projects that rethink the sustainability of our construction practice. "Tabula Scripta" introduces transformation projects exploring how we rebuild under pre-existing conditions. "After the NAi" presents a body of public projects built during the period after 2012 when the national budget was cut. Each theme is accompanied by an essay written by Bob Witman, Jarrik Ouburg, and Sergio M Figueiredo, respectively. Finally, Floris Alkemade, Chief Government Architect of the Netherlands, shares with us in an interview about his reflections, actions, and predictions on the country's architectural setting.

Credits to Eelco Van Welie, Director of nai010 publishers, for his support, including authors of the essays, and the architects who contributed to this book. (a+u)

在本书中，我们邀请到nai010《荷兰建筑年鉴》的总编克丝汀·汉内马担当客座主编，与我们一起回顾了荷兰建筑近十年的发展历程。a+u编辑的上一本荷兰专辑还是出版于2012年1月的《a+u 12:01 荷兰建筑 2000-2011》，当时编辑那本专辑的时候，也许正值2008年全球经济危机后不久，建筑业一片衰颓，但那时"典型的"荷兰建筑已初显端倪。试将我们对荷兰建筑的印象从"超级荷兰"的标签上移开，就会看到结合了更广泛的社会问题、具有新文化力量的项目。然而，对比上一代的超级巨星，这些建筑师和他们的作品仍常被忽视。因此，本书精选了19个项目，将分三个主题，向大家介绍萌芽的荷兰当代建筑新潮——"超常规"。

三个主题分别为："重塑围垦地"，通过一些实验性作品，重新思考建设行为的可持续性。"Tabula Scripta"，通过介绍一些改建、改造的作品，拷问既存建筑被再建筑的可能性。第三部分是"NAi以后"，主要介绍了2012年荷兰建筑协会(NAi)的预算被削减后建设的一系列公共项目。每一个主题都附有一篇论文，分别由鲍勃·惠特曼、贾里克·奥伯格和塞尔吉奥·M.菲格雷多撰写。书的最后，荷兰的首席政府建筑师弗洛里斯·阿尔马德在一次采访中与我们分享了他对荷兰建筑背景的思考、行动和预测。

感谢nai010出版社的负责人伊尔科·范·韦利的支持，包括论文的作者，以及参与本书编辑的所有建筑师。

(a+u)

Introduction:
SuperNormal
Kirsten Hannema

导读：
超常规
克丝汀·汉内马

The request to look back on Dutch architectural production of the past decade for this edition of *a+u* involuntarily brought me back to November 9, 2007. On that day, the Netherlands Architecture Institute (NAi) organized the symposium *Architecture 2.0 – The Destiny of Architecture*. At the invitation of NAi-director Ole Bouman, 6 Dutch architects – WielArets, Ben van Berkel, Francine Houben, Rem Koolhaas, Winy Maas and Willem Jan Neutelings, known as the generation SuperDutch – came together in the architectural city of Rotterdam to discuss the future of Dutch architecture.

Apparently architecture was doing well. Since SuperDutch – named after the book that architecture critic Bart Lootsma wrote in 2000, about how this group of architects conquered the world with their radical concepts and spectacular buildings in the late nineties – the Netherlands had become the mecca of modern architecture and this architecture an extremely successful export product. But something had started to gnaw.

Bouman first of all expressed his doubts about the spatial policy of Rotterdam, which seemed primarily focused on building iconic projects. In his preliminary statement, he also noted that, "The architect no longer has the same position in the building process as before. Various forces, much stronger than any talent or genius, influence the work of the architect. Financing models, ownership relationships, globalization, new media, legalization..." But he suspected that the SuperDutch architects, because of their enormous international success, "held a key to the capacity of architecture."

Willem Jan Neutelings' presentation was the one that stood out. He stated that, "The most important task for the next twenty years is not to design megalomaniac monuments for the new Sun Kings of this world in Dubai or Beijing, but to design a comfortable daily living environment for billions of people around the world. The task lies in the design of the everyday."

His words caused confusion and scorn among the architects in the audience. What was his plan? That he and his SuperDutch-colleagues would stop building theaters and museums? That they would go "back to basics" and concentrate on designing flats and terraced houses? Did he mean that they should focus less on the architectural object and more on creating city ensembles, with pleasant streets and squares? Or was he saying that there should be more attention for the architects who were involved with such banal tasks, but for whom there was hardly any attention in the media – and who were not invited to speak at the symposium.

Of course Neutelings didn't have a plan. But 13 years – and one global economic crisis – later we may conclude that all this more or less happened.

No, Neutelings Riedijk Architects did not stop building culture palaces. But their project for the Deventer City Hall (see pp. 110–121) illustrates how they did take a new direction. In 2011, citizens organized protest marches against the project, calling the design with its giant cupola "megalomaniac". The architects were given the opportunity to adapt the design, and decided to insert the complex almost invisibly into the historic urban fabric. Looking for a way to "reconnect" the project to the public, they decorated the facades with an art work, made out of 2264 fingerprints of Deventer

得知要为这本《a+u》专辑撰写导读,来回顾荷兰过去十年间的建筑作品,我不禁回想起了2007年11月9日。那天,荷兰建筑协会(NAi)举办了"建筑2.0——建筑的宿命"研讨会。应该协会主任奥莱·布曼的邀请,6位"超级荷兰"一代的荷兰建筑师齐聚建筑之都鹿特丹,共同探讨荷兰建筑的未来,包括威尔·阿雷茨、本·范·伯克尔、法兰馨·侯班、雷姆·库哈斯、维尼·马斯、威廉姆·扬·诺特林斯。

显然,过去荷兰的建筑业发展得不错。建筑评论家巴特·卢茨玛在2000年的《超级荷兰》一书中,讲述了一群建筑师如何在1990年代末以激进的概念和壮观的建筑征服世界。从"超级荷兰"时期以来,荷兰也成了现代建筑的圣地,建筑服务的对外出口大获成功。然而事实上,有些东西已经开始被吞噬。

布曼率先对鹿特丹的空间政策提出了质疑,建设地标项目似乎成了工作重心。他在活动的开场白中指出:"建筑师在建造过程中的地位大不如前。远比天赋或才华更为强势的多方力量正在影响建筑师的工作,例如金融模型、股权结构、全球化、新媒体、司法化……"但他也认为,"超级荷兰"的建筑师正是由于在国际上取得了巨大的成功,才"拿到了一把打开建筑潜能的钥匙"。

威廉姆·扬·诺特林斯的发言则格外引人注目。他表示:"未来二十年最重要的任务,不是为迪拜或北京这种现代'太阳城'设计巨型纪念碑,而是为全球数十亿人设计舒适的日常生活环境。挑战潜藏于对日常的设计之中。"

他的话引起了在场建筑师们的困惑和不屑。他的企图是什么?难道他和他"超级荷兰"的同事们将停止建造剧院和博物馆?难道他们要"返璞归真"、专注设计公寓和排屋?他是认为,建筑师应该把建筑单体的关注转移到城市整体上,创造更多令人愉悦的街道和广场?还是说,那些参与这类平凡工作、极少被媒体聚焦、甚至没有受邀在研讨会上发声的建筑师,应当获得更多的关注?

诺特林斯当然没有企图。但13年(其间还发生了一次全球经济危机)之后,我们可能会得出结论:这一切或多或少发生了。

诚然,诺特林斯所在的诺特林斯-里迪克建筑师事务所没有停止建造文化殿堂,但他们设计的代芬特尔市政厅(见110-121页)说明了他们是如何踏上新道路的。2011年,市民组织抗议游行,谴责市政厅带有巨型穹顶的设计"浮夸狂妄"。因此,在得到修改的机会后,建筑师决定将整个建筑群温和地嵌入古老的城市肌理之中。他们还在建筑外立面设计了一件由2,264个市民指纹组成的铝制艺术作品,试图"重新连接"市政厅与公众的关系。

2014年,雷姆·库哈斯"返璞归真",担任威尼斯建筑双年展的主策展人。当时的展览主题是"建筑元素",因而他与哈佛大学的学生一起调查了窗户、阳台、卫生间之类的普通事物,展示了历代建筑师如何利用它们创造出非凡的空间,激励新一代建筑师再次效仿。

另一个化腐朽为神奇的例子,体现在2014/2015年荷兰建筑年鉴的双封面上。正面是MVRDV建筑师事务所的鹿特丹拱廊市场(a+u17:04),背面是阿姆斯特丹的德·苏维尔(见48-57页)。后者是一个DIY项目,废弃船屋摇身一变,成了艺术家的工作室。两年后的2017年,位于阿姆斯特丹战后住宅区庇基莫尔的公寓改造项目——克莱堡公寓(De Flat,见176-183页),还获得了大名鼎鼎的密斯·凡·德·罗奖。历届政府建筑师都操刀了"宏大项目",比如荷兰国立博物馆的改造(见232-245页,经过十年改造后于2013年重新开放),而现任政府建筑师弗洛里斯·阿尔克马德(OMA的前合伙人)却在从事"普通"工

citizens, cast in aluminium.

In 2014, Rem Koolhaas went "back to basics" as director of the Architecture Biennale in Venice *Elements of Architecture*. Together with Harvard students he investigated ordinary things like the window, the balcony and the toilet, demonstrating how architects over time have used them to create extraordinary spaces and thereby inspiring the new generation to do so again.

Another example of how banal became "the new black" was the double cover of the Yearbook Architecture in The Netherlands 2014/15, with MVRDV's Rotterdam Martkhal (*a+u* 17:04) on the front and on the back De Ceuvel (see pp. 48–57) in Amsterdam, a DIY-project comprising studio spaces built with discarded houseboats. 2 years later De Flat (see pp. 176–183), a renovated gallery flat in the post-war district of Amsterdam Bijlmermeer, won the prestigious Mies van der Rohe Award 2017. And while previous government architects rigged up Grand Projects, such as the renovation of the Rijksmuseum (see pp. 232–245, reopened in 2013 after 10 years of renovation), the current government architect Floris Alkemade – former partner at OMA – organizes competitions for "ordinary" tasks such as affordable housing and the renewal of the countryside.

There it is in a nutshell, how a different way of architectural thinking and a new energy came into the Netherlands – this we might call SuperNormal. But, how did that happen?

Architecture 2.0 showed that Dutch architecture was in a substantive crisis already before the global economic crisis struck and turned everything upside down. Neutelings speech might have evoked a spark, the fire started in 2008 when Lehman Brothers fell over. Construction projects were stopped, 40 percent of the architects lost their job. Brand new offices – built for speculation – were left vacant.

By now, the economy is booming again and new iconic buildings have emerged, such as the rock-shaped Forum Groningen by NL Architects (See pp. 192–199). A spectacular culture project that was started before the crisis, then much criticized because of the rising costs, but after all completed. Looking back, the crisis is seen as a difficult but also a fertile period, that offered time for reflection, and the experiment. After the SuperDutch era which focused on a conceptual way of designing, basic spatial qualities such as daylight and materiality again came to the fore, and the making itself. Asking how architecture could contribute to a pleasant living environment, everyday subjects such as good homes, a livable city and a climate-proof country became the new assignments. The subjects were perhaps more humble than in the SuperDutch period, but the ambition did not lie.

The crisis also has a dark side. In 2011, State Secretary for Education, Culture and Science, Halbe Zijlstra decided to cut back 200 million euros on culture; one fifth of the total budget. As a result, the NAi, which since its creation in 1988 had become the figurehead and "promotion machine" of Dutch architecture – merged into the New Institute in 2012. The Netherlands Architecture Fund merged into the Creative Industries fund; the new name making clear how the government, under the influence of neo-liberal policy, nowadays approaches architecture: as part of the economy.

作,例如设计经济适用房、区域更新等。

简而言之,一种截然不同的建筑思维方式和一股全新的能量来到了荷兰。我们可以称之为"超常规"。但是,这究竟是如何发生的?

建筑2.0表明,在全球经济危机袭来并颠覆一切之前,荷兰建筑就已经陷入了实质性危机。诺特林斯的演讲或许炸出了火花,但真正引燃变化的是2008年雷曼兄弟的破产。它直接导致了各地项目的停工,40%的建筑师失去工作,投机建造的全新办公室无人问津。

如今,经济回暖,新的标志性建筑再度出现,NL建筑师事务所设计的岩石状的格罗宁根广场(见192-199页)正是一例。这个令人惊叹的文化项目在危机到来之前开工,后来因成本增加而饱受诟病,但终究还是完成了。回首过往,尽管萧条时期困难重重,却给予了我们反思和实验的时间。在注重概念设计的"超级荷兰"时代之后,人们的焦点回到了自然光,物质材料等基本空间品质及其实现过程上。人们开始思考建筑如何推动创造一个愉悦的生活环境,设计主题变得更加日常,例如"舒适的家""宜居的城市""可应对气候灾害的国家"。与"超级荷兰"时代相比,这些主题或许有些低调,但雄心壮志清晰可见。

这场危机同样有黑暗的一面。2011年,荷兰教科文部国务秘书哈尔伯·泽尔斯特拉决定削减两亿欧元的文化经费,这相当于总预算的五分之一。于是,1988年成立、作为荷兰建筑形象代言和"推广机器"的NAi,在2012年被并入了新的机构;"荷兰建筑基金"则被并入了"创意产业基金"。新名称显示出,在当今新自由主义的影响下,政府将建筑作为经济的一部分。更早的一年前,荷兰还撤销了住房、空间规划和环境部,工作分别由基础设施和环境部、内政部以及经济事务、农业和创新部接管。1990年代刺激建筑业发展的"空间规划""空间规划备忘录"等名词,已从官方政策中消失。总而言之,今天荷兰的建筑环境不容乐观。

面对经济的萧条,人们纷纷走上不同的道路,多元化成为当下建筑界的一大特点。超现代主义者、后现代主义者、理性主义者、空间研究者、专业修缮建筑师、景观设计师、城市设计师、室内建筑师、艺术家和DIY爱好者开始并肩同行。尽管工作类型和方式千差万别,我们仍可以从中看出三大趋势。

首先,"超级荷兰"建筑师仍有一席之地。包括KAAN、本特姆·克劳威尔在内的大型事务所继续受到资本的青睐,他们有机会操刀国外的项目(经济危机期间这样的机会更多)以及本国的项目,其中不乏气势恢宏的图书馆和剧院,但大约从"建筑元素"双年展开始,出现了"常规"的倾向。麦肯诺建筑师事务所正在对纽约公共图书馆和华盛顿的马丁·路德·金纪念图书馆进行适度改造;OMA在鹿特丹建造了"反标志性"的市政大楼Timmerhuis(a+u 15:09),看上去如同一朵溶于城市中心的"像素云";科恩·范·维尔森设计的布雷达汽车站与火车站(见220-231页)遵循同样的理念,铁轨之上的巨大砖造综合体堪比一座小城,容纳了住宅、办公室、商店、停车场等,重新连接起轨道两侧的街区。

其次是"新荷兰"。瑞士建筑杂志《Modulor》曾经用这一名称介绍了以汉斯·凡德海登、Winhov事务所、Monadnock建筑师事务所、哈佩尔–科尼利斯–韦尔霍芬建筑师事务所为中心的建筑师群体。他们的作品里,文脉、传统、工艺和材料占据了中心地位。这种反"超级荷兰"的运动并不新鲜,自现代主义兴起以来,荷兰的现代派和传统派就没有停止过"对战"。但不同的是,与十年前

The Ministry of Housing, Spatial Planning and the Environment was abolished a year earlier; its tasks were assigned to the Ministries of Infrastructure and the Environment, the Ministry of the Interior and the Ministry of Economic Affairs, Agriculture and Innovation. The term Spatial Planning and Spatial Planning Memorandums with which architecture was stimulated in the 1990s disappeared from government policy. All in all, the architectural climate in the Netherlands is harsh today.

In times of crisis, everyone runs in a different direction. Diversity characterizes the current architectural landscape, in which supermodernists, postmodernists, rationalists, spatial agencies, specialized restoration architects, landscape designers, urbanists, interior architects, artists and do-it-yourselfers work alongside and with each other. Yet, from these many office types and styles three trends can be discerned.

To begin with, the SuperDutch architects are still there: the large offices, to which we may also count KAAN and Benthem Crouwel. Their portfolios are well stocked, with projects abroad – where there were more opportunities during the crisis – but also in their own country. Among these projects are spectacular libraries and theaters, but roughly since the *Elements of Architecture* biennial we have also seen a more "normal" side. Mecanoo is currently working on the modest renovation of the New York Public Library and the Martin Luther King Jr. Memorial Library in Washington. OMA built the "anti-iconic" Stadstimmerhuis (*a+u* 15:09) in Rotterdam, a "pixelated cloud" that "dissolves" in the inner city. Koen van Velsen's design for Breda's railway station (see pp. 220–231) follows the same line; the gigantic brick complex – with houses, offices, shops and a parking garage – a small city in itself - has been built over the tracks, reconnecting the neighborhoods on either side of the railway.

Then there's the *New Dutch*, as the Swiss magazine Modulor presented the group of architects around Hans van der Heijden, Office Winhov, Monadnock and Happel Cornelisse Verhoeven. In their work, the context, tradition, craftsmanship and materials take the central stage. This counter movement of SuperDutch is not new. Ever since the rise of modernism there has been a "battle" between modernists and traditionalists in the Netherlands. The difference is that there is now much more attention for this everyday architecture, as compared to 10 years ago. This has to do not only with the changing sentiment during the crisis, when extravagance was replaced with sobriety, but also with the nature of the current building assignments.

Partly as a result of the enormous vacancy of buildings and the orientation towards sustainability, a shift has taken place in recent years from expansion to urban densification, from new construction to transformation. *Tabula scripta*, as Floris Alkemade calls this new condition, which he made into the theme of his research group at the Amsterdam Academy of Architecture in 2014. It is a condition that requires a different attitude and approach than what was used in the past recent decades: from drawing on a blank sheet of paper to building on the existing one. They are the kind of tasks in which these architects feel at home.

相比,如今人们对日常建筑的关注度大大提高。这一方面是因为萧条时期舆论倾向由铺张转为节制,另一方面也与当前建筑的本质有关。

近年来,随着建筑的大规模空置,人们的环保意识不断增强,城市逐渐从扩张变成收缩、密集,建筑更多源于改造而非新建。弗洛里斯·阿尔克马德将这种新状况称为Tabula Scripta(译注:研究名词,可理解为"刻有文字的石板"或"有脚本"),2014年时还把它列为阿姆斯特丹建筑学院研究小组的课题。此时,过去几十年的态度和方法都不再奏效,因为原先是在白纸上从无到有地创作,如今却是在已经存在的事物上建造。但对于"新荷兰"建筑师而言,这些任务能让他们感到自在。

第三种趋势,是朝着全新的方向,重塑建筑师的角色和建造的流程。不少团队在危机到来之前就开始活跃,包括专注物流的超级利用工作室。另一些则因为没有任何建筑委托,在萧条时期"被迫"进行新的尝试,例如参与小规模的DIY项目、利用废旧材料、使用3D打印等新技术。而随着开发商和企业不再有能力为大型项目提供资金,市政当局不得不依靠个人投资建造新楼,一些建筑师和小型开发商抓住机会启动了新项目"Baugruppen"。这是一种简单(且经济)的双层复式公寓,住户可以按照自己的意愿装修内部。在早有实验性建筑项目的阿尔梅勒,还出现了新型的自建社区Oosterwold(由MVRDV总体规划),居民不仅自己建房,还负责铺设街道。

本书选取的19个项目正是按照这三种趋势分类。

RAAAF,里昂工作室在项目"三角洲之作//"中(见76-83页),将用于测试东谢尔德大坝(三角洲工程的一部分)的等比例模型Deltagoot从中锯开,他们想借此提出问题:荷兰一贯的造坝围垦模式是否仍然可持续?除此之外,还有一系列先锋项目,关注着新型居住模式、遗产应对策略、气候变化等课题。建筑记者鲍勃·惠特曼一直在观察这些"原型",并对话了一些景观建筑师。这些建筑师给荷兰提出了新的规划方案,比如建造人工沙丘,将围垦区"还于"自然。

建筑师贾里克·奥伯格是阿姆斯特丹建筑学院阿尔克马德研究小组的成员,这个小组关注设计实践在Tabula Scripta状态下的变化。他们认为文脉不仅不是限制,反而还提供了挖掘场所潜力的机会。他们自问:怎样重新解释、理解、尊重和发展当前的环境?遗产保护和开发怎样才能携手并进?在此基础上,建筑又可以期待怎样的新发展?

塞尔吉奥·M·菲格雷多是《NAi效应——创造建筑文化》(2016)的作者,本书有幸邀请到他撰稿,结合政治和建筑环境,对NAi关闭后的"宏大项目"进行审视。

尽管本书的焦点是荷兰当代的建筑进展,但对思考我们今后的道路仍然至关重要。为此,我采访了政府建筑师弗洛里斯·阿尔克马德,请他分享如何看待荷兰建筑乃至这个国家的未来。当我让他用一句话描述未来十年时,他重复了诺特林斯在2007年的言论:下一个超级将指向"超泛化"。

Thirdly, there is the group who took a complete new direction, reinventing the role of the architect and the building process. A number of these agencies were already active before the crisis, such as Superuse Studios, specialized in connecting material flows. Others were "forced" to do something else during the crisis, as they didn't have any building commissions, and started small scale DIY-projects, experimenting with waste materials and new technologies such as 3D-printers. Because developers and corporations were no longer able to finance large scale projects, municipalities had to rely on private individuals for the construction of new homes. Some architects and small developers saw an opportunity and started new projects with so-called *Baugruppen*. They designed simple (and cheap) double-height lofts in which the residents can build their interior according to their wishes. In Almere, which has a long history of experimental projects, a start was made with the construction of the new self-built neighborhood Oosterwold (master-planned by MVRDV), where residents not only build their own houses, but also construct the streets.

The 19 projects selected for this book are grouped along these 3 lines.

With their project Deltawerk// for which RAAAF and Atelier de Lyon (See pp. 76–83) put the saw in the Deltagoot, a 1:1 model which was used to test techniques for the Oosterscheldekering (part of the famous Delta Works) they ask whether the Dutch polder model with flood defenses and dikes still is sustainable. It is one of a series of pioneering projects focusing on new forms of housing, heritage strategies and climate change. Architecture journalist Bob Witman examined these "prototypes" and entered into discussions with landscape architects who propose radically new planning strategies for the Netherlands, such as the construction of artificial dunes, while existing polders are "returned" to nature.

Architect Jarrik Ouburg, Fellow of Alkemade's research group at the Amsterdam Academy of Architecture, focuses on new design practices working within the *Tabula Scripta* condition. They see the context not as a restriction, but as an opportunity to exploit the potential of the place, asking themselves: how can the existing context be reinterpreted, understood, appreciated and further developed? How can the concern for heritage and new developments go hand in hand? And how can architecture anticipate new developments based on the existing situation?

Sergio M. Figueiredo, author of *The NAi Effect-creating architecture culture* (2016), was invited to look into the Grand Projects, in relation to the changed political and architectural climate, after the closure of the NAi.

While this issue focuses on the current progress of architecture in the Netherlands, it is vital to think about the upcoming course we are going to take. For this, I spoke with Government Architect Floris Alkemade to have him share with us his thoughts on the direction Dutch architecture has taken and his stand on the country's future. When I asked him to describe the upcoming decade in one sentence, he repeated what Neutelings said in 2007: the next super will be the generic.

Kirsten Hannema studied architecture at Delft Technical University and is currently working as a freelance architecture critic, among others, for the Dutch newspaper, *De Volkskrant*. She has worked as an editor for the architecture magazine *A10 European Architecture* from 2005 to 2012, and her articles have appeared in international magazines. Her aim is to write not only for architectural insiders but also for a wider audience, bringing relevant social topics to the table, in an understandable language and attractive articles. She has worked on publications such as the *A10 New European Architecture Yearbooks*. She was a jury member for numerous architecture prizes and guest teacher at TU Delft and the Amsterdam Academy of Architecture. Since 2016, she is editor in chief of the *Yearbook Architecture in the Netherlands*.

克丝汀·汉内马曾在代尔夫特理工大学学习建筑，目前是建筑方面的自由评论家，为荷兰《人民报》等媒体供稿。她曾在2005年至2012年以编辑身份就职于建筑杂志《A10欧洲建筑》，她的文章常在国际杂志上发表。她的目标是不仅要为建筑业内人士写作，也面向更广泛的读者撰写通俗易懂、精彩有趣的文章，引入相关的社会话题。她曾参与《A10新欧洲建筑年鉴》等出版物的创造。她还是众多建筑奖项的评委，也是代尔夫特理工大学和阿姆斯特丹建筑学院的客座教师。她自2016年起负责主编《荷兰建筑年鉴》。

the Netherlands
荷兰

This map charts the locations of both the projects from a+u 12:01 (in grey), and the projects featured in this book (in orange). Refer to the visual chronology on pp. 21–29.

这张图显示了收录于《a+u 12:01》中项目的位置（灰色），以及本书收录项目的位置（橙色）。具体可以参考第21-29页的年表。

Selected Works in the Netherland (2020-2000)
荷兰建筑作品精选（2020～2000年）

1. Forum Groningen
NL Architects
Groningen, 2019
(pp. 192–199)

2. Museum De Lakenhal
Happel Cornelisse Verhoeven
Leiden, 2019
(pp. 166–175)

3. Naturalis Biodiversity Center
Neutelings Riedijk Architects
Leiden, 2019
(pp. 134–143)

4. Tij Observatory
RAU Architects, RO&AD Architects
Stellendam, 2019
(pp. 68–75)

5. Utrecht Bicycle Parking
Ector Hoogstad Architects
Utrecht, 2019
(pp. 200–207)

6. Deltawerk//
RAAAF, Atelier de Lyon
Marknesse, 2018
(pp. 76–83)

7. LocHal Library
CIVIC Architects
Tilburg, 2018
(pp. 122–133)

8. Musis Sacrum
Frits van Dongen Architects and Planners,
Koschuch Architects Arnhem, 2018
(pp. 156–165)

9. The Sixteen-Oak Barn
Hilberinkbosch Architects
Berlicum, 2018
(pp. 58–67)

10. BlueCity Rotterdam
Superuse Studios
Rotterdam, 2017
(pp. 102–109)

11. Co-living Oosterwold
bureau SLA, ZakenMaker
Almere, 2017
(pp. 38–47)

12. Delft City Hall and Train Station
Mecanoo
Delft, 2017
(pp. 208–219)

13. Beurspassage – Oersoep
Robert A.M. Stern Architects (RAMSA), Rijnboutt
Amsterdam, 2016
(pp. 94–101)

14. Deventer City Hall
Neutelings Riedijk Architects
Deventer, 2016
(pp. 110–121)

15. Kleiburg De Flat
NL Architects, XVW Architecture
Amsterdam-Zuidoost, 2016
(pp. 176–183)

16. Public Transport Terminal Breda
Koen van Velzen Architects
Breda, 2016
(pp. 220–231)

17. De Ceuvel
Space&Matter
Amsterdam, 2014
(pp. 48–57)

18. Education Center Erasmus MC
KAAN Architects
Rotterdam, 2013
(pp. 144–155)

19. The Rijksmuseum
Cruz y Ortiz Architects
Amsterdam, 2013
(pp. 232–245)

20. Inntel Hotels Amsterdam-Zaandam
WAM architects
Amsterdam-Zaandam, 2010

21. Knikflats
Biq Stadsontwerp
Rotterdam, 2010

22. Recycle Office – HAKA building
Doepel Strijkers Architects
Rotterdam, 2010

23. Hiphouse
Atelier Kempe Thill
Zwolle, 2009

24. Le Medi
Geurst & Schulze Architects
Rotterdam, 2009

25. Office and Architect's Dwelling on Boomgaardsstraat
Kühne & Co
Rotterdam, 2009

26. Rehabilitation Clinic Groot Klimmendaal
Koen van Velsen
Arnhem, 2009

27. Schieblock
Zones Urbaines Sensibles
Rotterdam, 2009

28. Villa Welpeloo
2012Architects
Enschede, 2009

29. CitizenM Hotel
Concrete
Amsterdam, 2008

30. La Grande Cour
Meyer en Van Schooten Architects (MVSA)
Amsterdam, 2008

31. Rebuilding Roombeek
CPi de Bruijn, de Architects Cie.
Enschede, 2008

32. Bureau IJburg
Claus en Kaan
Amsterdam, 2007

33. De Matrix Community School
Architect bureau Marlies Rohmer
Hardenberg, 2007

34. Klavertje 4
Studio Marco Vermeulen
Venlo, 2007

35. NDSM Studio City
Dynamo Architects
Amsterdam, 2007

36. Wallisblok
Hulshof Architects
Rotterdam, 2007

37. Westraven
Cepezed Architects
Utrecht, 2007

38. Netherlands Institute for Sound and Vision
Neutelings Riedijk Architects
Hilversum, 2006

39. Ypenburg Center
Rapp + Rapp
The Hague, 2006

40. Castle Leliënhuyze
Soeters Van Eldonk Architects
's-Hertogenbosch, 2005

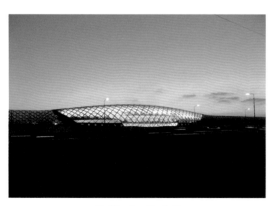

41. Hessing Cockpit
ONL
Utrecht, 2005

42. Royal Netherlands Embassy
Dick van Gameren and Bjarne Mastenbroek
Addis Ababa, Ethiopia, 2005

43. WKK Energy Plant
Liesbeth van der Pol, Dok Architects
Utrecht, 2005

44. Parkrand, Osdorp
De Nijl
Amsterdam, 2004

45. Souterrain Tram Tunnel
OMA
The Hague, 2004

46. University Library, Utrecht
Wiel Arets
Utrecht, 2004

47. The BasketBar
NL Architects
Utrecht, 2003

48. De Resident
Rob Krier · Christoph Kohl · Architects
The Hague, 2001

49. Hageneiland
MVRDV
The Hague, 2001

Credits for Photographs on pp. 20–29
All photos, except as noted, courtesy of the architects.
1. by Katja Effting, 2., 15. by Scagliola Brakkee, 3. by Karin Borghouts, 4. by Stijn Bollaert, 5. by Marcel van der Burg, 6. by Petra Appelhof, 7., 13. by René de Wit, 8. by Bart van Hoek, 9. by Jan Kempenaers, 10. by Filip Dujardin, 11. by Mecanoo architects, 12. by Frank Hanswijk, 14. by Marcel van der Burg, 16. by Kees Hummel, 17. by Fernando Guerra FG+SG, 19. by Duccio Malagamba, 20. by Ralph Kämena, 21. by Stefan Müller, 22. by Roel Backaert, 23. by Allard van der Hoek, 25., 48. by Rob 't Hart, 26. by Stefan Müller, 27. by Ulrich Schwarz, 28. by Edwin Prins, 29. by Irvin van Hemert, 30. by Ewout Huibers, 31., 32., 34. by Jeroen Musch, 33. by Frank Hanswijk, 37., 38. by Kim Zwarts, 39., 43 by Daria Scagliola & Stijn Brakkee, 40., 45. by Hans Werlemann, 42. by Rob hoekstra, 44. by Jan Bitter, 46. by Jannes Linders, 47. by Hans van Leeuwen, 49. by Mick Palarczyk.

Projects 1–19 are featured in this book. Projects 20–49 were featured in a+u 12:01, Architecture in the Netherlands 2000–2011.

本书专题介绍项目 1–19。项目 20–49 收录在《a+u 12:01 2000-2011 年荷兰建筑》中。

Essay:
Reshaping the Polder
Bob Witman

论文：
重塑围垦地
鲍勃·惠特曼

A giant egg made from sustainable softwood, reed and rocks lies on a mud plate near the imposing key piece of the Dutch Delta Works: the Haringvliet sluices (1958–71). The contrast between the egg – which serves as a bird observatory – and the flood barrier could not be bigger. Opposite the striking concrete, steel and basalt construction to keep the high seas out, RAU Architects and RO&AD Architects installed a delicate wooden building that invites nature to come in. The sanctuary offers a good view of the quail and spoonbills that forage in the tidal landscape.

This Tij Observatory (2019, see pp. 68–75) has been set up to celebrate the Haringvliet sluices that allow salt water from the North Sea to pass freely. This was desperately needed, because the flood barrier had turned the inland waterways into a dead brackish ecosystem. The original flora and fauna were totally destroyed. Now a constant stream of sea water flows in through a crack in the dam, and nature restores itself as if a new springtide has arrived.

With the Dutch polder changing, a place in the province of Zeeland can be seen as a symbol where larger movements are taking place. The long-standing Dutch approach of keeping water out with hard walls (dikes) has turned to moving with nature's flow. It manifests in the choice of materials used in architecture and the liberation of the landscape – where some polders must make way for water as rivers are allowed to overflow. Especially by the end of the Vinexwijk, it shows that the typical neatly structured Dutch polder towns are dismissed in favour of a smaller-scale sustainable approach, directed towards DIY-construction, unkempt nature, and with lesser rules to follow.

Moving along with nature's flow is not just an ecological necessity, it is fiercely inspired by pragmatic arguments. The densely populated Randstad in the west part of the Netherlands, manages more than half of the economic capital – 400 billion euros per year in GNP value. Coincidentally, this financial engine – that houses half of the population – is mainly situated in polders that lies several meters below sea level [we are a people of thrill seekers]. If all that economic capital washes away, little wealth would remain in the Netherlands.

"In the long term, you have to recognize the possibility that the rise of sea level can not be tackled with higher dikes and heavier pump stations. That it is not sustainable. There is a limit to the idea of the good old polder," says landscape architect Kristian Koreman of ZUS (Zones Urbaines Sensibles) in Rotterdam. Due to climate change, the sea can rise up to 4 or 5 meters above our polder level in the next century. "We have to think of a bigger escape-plan."

Pinpricks from that big escape that ZUS has in mind can be found in Almere – the youngest city in the Netherlands (1975), located in the Flevopolder. In the outskirts of this town ZUS designed a new residential area called Almere Duin (see p. 34), a friendly neighbourhood with brick houses that lie between the marram grass and rolling shell paths. It is a beach landscape where you can almost smell the salty sea. But there is no salt, no sea. The houses are scattered on high sand dunes that are completely artificial. "Yet, you have the idea that you are constantly on holiday here," one of the residents once said.

For ZUS, Almere Duin is more than just a holiday feeling. It is their impetus for a visionary delta plan that must keep Dutch feet dry. There is more than enough sand near the coast of the – shallow

哈灵水道挡潮闸(1958-1971年,荷三三用洲工程的关键部分)附近的泥板上,可以看到一个由可持续利用的软木、苦苇和岩石构成的巨蛋,这是一座鸟类观测站。防洪坝与巨蛋之间形成了鲜明的对比:前者是混合了混凝土、钢铁、玄武岩的大型结构体,将高水位的海潮隔挡在外;与之相望的后者则是一座由RAU和RO&AD两家建筑事务所共同设计的木建筑,外观别致,仿佛向自然正发出邀请。从巨蛋中望出去,可以很好地观赏到在潮池中觅食的鹬鹬和琵鹭。

这座Tij鸟类观测站(2019年,见第68-75页)的设立,是为了庆祝哈灵水道挡潮闸的开闸。由于防洪坝已经将内陆水道变成一个死气沉沉的咸水生态系统,原先的动植物完全被破坏,开闸让北海盐水自由通过就显得极其必要。现在,海水可以从大坝的缝隙中源源不断流入,大自然终于恢复原貌,仿佛迎来了新一波的春潮。

荷兰围垦地的变化,在进行大规模运动的泽兰省尤为明显。长久以来,荷兰使用硬墙(堤坝)挡水,现在却开始顺应这种自然的流动。通过建筑材料的选择和景观的解放可以看到,部分围垦地必须给水让路,以便河流溢出。特别到了VINEX(第四次国土规划修正案)时,循规蹈矩的围垦造城方式被摒弃,小规模、可持续的开发受到欢迎;后者采用直接手工建造,面向本真的自然,没有那么多的条条框框。

除了出于生态的需要,顺应自然流动也是受到实用论的刺激。荷兰西部的兰德斯塔德市人口密集,掌握着全国半数以上的经济资本,每年国民生产总值达到4,000亿欧元(约合人民币31059.6亿元)。巧合的是,这个容纳了荷兰一半人口的金融引擎大部分坐落在低于海平面数米的围垦地(我们是一个追求刺激的民族)。如果这些经济资本被全部冲走,荷兰的财富就将所剩无几。

鹿特丹ZUS(Zones Urbaines Sensibles)事务所的景观建筑师克里斯蒂安·科雷曼坦言,"你必须认识到,从长远来看,海平面上升的问题不可能依靠更高的堤坝和更重的泵站来解决。那是不可持续的。传统围垦理念的可取性很有限"。由于气候变化,在下个世纪,海水可能会继续上升,最终高出现在安全堤的4米或5米。"我们必须想出一个更大的逃生计划。"

ZUS逃生计划的雏形可以在弗莱福兰省的阿尔默勒,荷兰最年轻的城市(建于1975年),找到。ZUS在郊区设计了一个氛围友好的新住宅区阿尔默勒·杜因(见第34页),砖造房屋坐落在滨草和蜿蜒的贝壳小径之间,你甚至仿佛能够闻到海潮的清香。然而,那里没有盐,没有海,房屋散落在完全人造的高沙丘上。"尽管如此,你会感觉正在这儿度假",一位居民曾这样说。

对于ZUS来说,阿尔默勒·杜因不只是为了打造度假的感觉,这更是他们进行三角洲计划的动力,他们希望借此让荷兰人立足于干燥之地。科雷曼说,在北海的浅海海岸附近,有足够多的沙子创造高地。你可以从海里取出240亿立方米的沙子,用人工沙丘抬高所有低矮的围垦地,从而保障人们的安全。

这一愿景最终被ZUS融入了Duinmetropool(沙丘大都市计划,见第33页),以2150年荷兰地图的形式呈现。"巨大的沙丘区域将覆盖几乎半个荷兰,同时,这里也将成为世界上最大的天然的水处理厂。"科雷曼说,"没有比通过沙子过滤雨水更好的提取饮用水的办法了。这是一个非常可持续的过程——不需要任何能源。"

当然,沙丘大都市计划还是一个遥远的设想。但是人们已经意识到,围垦地必须为一个全新的未来做好准备,这也体现在了许多小型项目之中。荷兰西海岸一个古老的

– North Sea to create high ground, says Koreman. You could lift 24 billion m3 of sand from the sea and raise all low polder areas with artificial dunes so that we are safe.

This scenario was coined by ZUS as the Duinmetropool (Dune Metropolis, see p. 33), taking shape as a map of the Netherlands in 2150. "That enormous dune area that will cover almost half of the Netherlands is, at the same time, also the largest natural water treatment plant in the world," says Koreman. "No better way to filter drinking water from caught rain water through the dune sand. A very sustainable process – no energy needed."

The Duinmetropool is of course a far-away scenario. But the idea that the polder must already prepare itself for a new future is reflected in many smaller projects as well. Katwijk, an old fishing village on the west coast of the Netherlands, was on the verge of watching the sea disappear behind an old-fashioned Dutch high dike because the natural coastal strip offers too little protection against rising sea levels that threatens the polders in the inlands.

OKRA Landscape Architects calculated that if you would spread 2.5 million m^3 of sand, a 100-meter deep dune landscape would be created that would offer just as much protection for Katwijk as high dikes. Now the sight lines from the village of the sea are preserved. And an underground parking lot was built into the dune, which earned the project the prize for the Royal Institute of Dutch Architects (BNA) Building of the Year 2016 in the Netherlands. A remarkable victory if you take into account that there is no building to be seen.

This appreciation for landscape architecture says a lot about how much careful consideration is given to the relationship between the natural landscape and its built environment in the Netherlands. The new polderprojects literally use what the country offers. For instance, The Sixteen Oak Barn (Hilberink Bosch Architects, see pp. 58–67) in the small town of Blaricum, that is nominated as BNA Building of the Year 2019. Here, it is not the aesthetic vision of the designer that led the way, but the application of the material sourced that did.

Every splinter of the sick old oaks on the courtyard of the building site was used to make the barn house, including the bark of the trees that has been poured into the concrete to let the building blend into its environment. The Sixteen Oak Barn is an example of this new "natural" aesthetic, in which visual choices are selected based on the availability of a material. It is a form of incidental aesthetic that can be said to be typical of Dutch sustainability.

This form of incidental aesthetic is also displayed in the "Commune-house" in the province of Drenthe made out of loam and straw, the best available local material, which was designed by Cesare Peeren. This is the architect who co-founded the design office Superuse Studios and is responsible for the circular BlueCity Rotterdam project (see pp. 102–109). His office also created Harvest Maps (oogstkaart.nl), an atlas to upload circular building material, so that designers can find nearby used material for their projects.

The choice of the shape in the new aesthetics are driven by the availability of the material, not by a notion of style. "I never start designing on a blank sheet," Peeren once explained in an interview, " I start with what is there on site, what is available." Sustainable architecture should be based on what you can reuse – harvesting on the existing like: sun

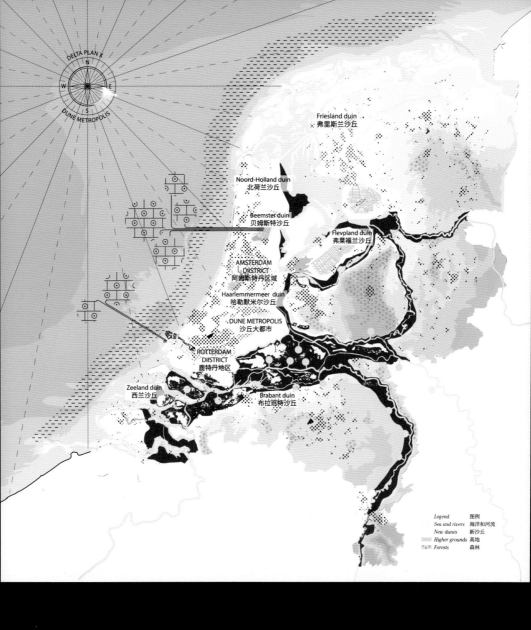

orientation, historical connotation, old building remains, and residual material from demolished constructions.

With regards to choice of materials, architects such as Thomas Rau – who designed the Tij Observatory – goes a step further. In his buildings, all the materials have their own passport. A kind of identity card that would guarantee how each material can receive a second life when life in its present building reaches an end.

De Ceuvel (see pp. 48–57) in Amsterdam-Noord is another project with that incidentally beautiful kind of identity. However, it is a success mainly because of its classic bottom-up origins. The self-built wooden constructions on the former shipyard exude a great DIY-atmosphere. There is a cafe-restaurant, a few small creative businesses and artists studios – all based together here in old houseboats that have been hoisted onto the dry banks. It also includes a hotel and a landing stage made of waste wood. Once shunned because it was on the "wrong side" of the IJ, today, this place has become one of the most trendy ventures in the booming north part of the city.

The history of De Ceuvel is deeply rooted in a real estate crisis. By 2012, there was no commercial developer to be found who wanted to invest in a neglected polluted shipyard. So, architect Wouter Valkenier and a group of idealistic entrepreneurs and artists acquired it for a period of 10 years. They turned it into a habitable "circular lab" during a

p. 33: Map of Dune Metropolis (Duinmetropool) – the Netherlands in 2150. This page, left: Aerial view of Almere Duin. Images courtesy of ZUS. Zones Urbaines Sensibles. This page, right: Photo along Almere Duin neighborhood with brick houses that lie between the marram grass and rolling shell paths. Photo by Walter Herfst.

第33页：沙丘大都市地图——2150年的荷兰。本页，左：俯瞰阿尔默勒·杜因；本页，右：阿尔默勒·杜因街区一带，砖造房屋坐落在滨草和起伏的贝壳小径之间。

对此,OKRA景观事务所给出了这样的计算结果:将250万立方米的沙子铺开,形成一个100米深的沙丘景观,同样能起到保护卡特维克(Katwijk)的作用。于是,这座村庄的海景被保留了下来。沙丘中建有一个地下停车场,还使得该项目获得了荷兰皇家建筑师协会(BNA)2016年度建筑奖。这是一次值得载入史册的胜利,毕竟从外观上,这里看不到任何建筑。

这种对景观建筑的认可,足以说明荷兰在处理自然景观和建筑环境的关系上有多么慎重。一些新的围垦地项目也表现出了对现有事物的利用,例如位于小镇布拉里克姆的十六号橡树谷仓(希尔伯林克·博世建筑师事务所,见第58-67页)。它还被提名了BNA 2019年度建筑,但原因不在于设计者的审美旨趣,而在于如何取材。

为了打造这座谷仓,连场地院子里奄奄一息的老橡树都被充分使用,树皮也被掺入混凝土中,彻底让建筑与环境融为一体。十六号橡树谷仓是典型的新"自然"美学,视觉呈现取决于可使用的材料。这种具有偶然性的美学形式堪称荷兰可持续发展的范例。

这一偶然美的形式同样表现在德伦特省的"公屋"中。该房屋使用了当地最容易获取的材料,黑土和稻草,并由塞萨雷·皮伦设计。这位建筑师是超级利用工作室的联合创始人,负责鹿特丹蓝色城市项目(见第102-109页)。他的工作室还上线了网站Harvest Maps(oogstkaart.nl),可以上传可循环建材的地理信息,便于设计师就近寻找材料。

这种全新的美的形态是源于可用的材料,而不是某种风格概念。"我从未在一张白纸上开始设计",皮伦曾在一次采访中解释道,"我从现场有什么、哪些可以使用开始"。可持续的建筑应当基于可循环的、已有的东西,例如阳光的朝向、历史的关联、旧建筑的遗存、拆解后的废弃材料。

在对材料的选择上,诸如托马斯·劳(设计了Tij鸟类观测站)等建筑师正在走得更远。他的建筑中,所有材料都有自己的"护照",如同身份证明一般,以此保证每种材料在当前建筑的使命终结时,能够获得第二次的生命。

位于阿姆斯特丹北的德·苏维尔(见第48-57页)是另一个具备这种偶然美的项目。然而,它的成功主要依赖其典型的、自下而上的发展。昔日的造船厂上,手工搭建的木建筑营造出浓郁的DIY氛围;被拖上岸的旧船屋里,集中着一家咖啡馆兼餐厅、一些小型创意店铺和艺术家的工作室。另外还有一间酒店和一个用废弃木头做的码头舞台。尽管这片区域曾因地处艾湾的"对岸"而被排斥,但如今已是快速崛起的城市北部最新潮的地点之一。

德·苏维尔的历史可以追溯到一次房地产危机。此前直到2012年,都没有任何企业开发商愿意投资这个被忽视、受污染的船厂。于是,建筑师伍特·沃克尼尔以及一群理想主义的企业家和艺术家买下了10年的使用权。在"循环建筑"一词还没有像今天这样被人接受的时候,他们就已经把它变成了一个适合居住的"循环实验室"。现在,德·苏维尔成了艾湾北岸的招牌,循环建筑项目在附近街区如雨后春笋般涌现。

"德·苏维尔的成功在于所有权的共享",沃克尼尔说。承担责任的不是投资者或地方政府,而是一群志同道合的理想主义者。不同领域的人分别负责修缮船屋、种植净化土壤的绿植、管理回收废弃物等,这才有了德·苏维尔。"作为建筑师,你肯定不是唯一的设计师。每个人都可以自己的方式添加一些东西。"德·苏维尔正是如此,过去

time when the term "circular building" was not as salonfähig (acceptable) as it is today. Now, De Ceuvel has become the poster boy of the north bank of the IJ and in the surrounding neighbourhood, circular building projects begin to spring up like mushrooms.

"The success of De Ceuvel lies in shared ownership," says Valkenier. It is not an investor or a local government who takes responsibility, but a group of like-minded individuals with an idealistic agenda. De Ceuvel relies on a diverse group of people to fix the old houseboats, to plant vegetation that extracts pollution from the soil, and to manage waste recycling. "As an architect you are certainly not the exclusive designer. Everyone adds something in their own fashion." And De Ceuvel is exactly that – an eclectically coloured mix of old materials and style, with an endless lively atmosphere which surprising, does very well on social media.

Perhaps, the financial crisis has not lasted long enough for these types of initiatives to reap the benefits from this slow development. After its recovery, land prices and pressure on the housing market continue to grow so much that time is now scarce in the real estate market of big Dutch cities. However, the notion that a real estate developer should take more patience in creating value still stands.

"It should be made clear to all of the investors that nature-inclusive building is in the short term more expensive, but it ultimately yields money," says Eric-Jan Pleijster, one of the 3 founders of LOLA Landscape Architects in Rotterdam. "We should look at the city as if it is a nature reserve. At this moment, a town like Rotterdam is one of the biggest polluters in the country," says Pleijster.

LOLA has developed a new toolbox to integrating nature in cities. The city and its buildings act as a wide range of ecotopes. Each ecotope can be enhanced by adding nesting possibilities for local species of insects, birds and bats. But above all, flora runs everywhere, along, up and through the buildings. Pleijster: "This way the experience of nature in the city is emphasized. And that's what makes people happy and healthy. Which in the end, also will be reflected in the value of the property."

This toolbox is part of a strategy which Pleijster calls Plan A. But when the effects of climate change aren't mitigated by the Paris Climate agreement, there is a need for Plan B. "In this Plan B for the year 2200, the Dutch will move along with what nature dictates. If the sea levels rise rapidly, we will have to give back most of the polders to the sea and river delta. We might be able to keep some historical city centers in the west," says Pleijster. But eastwards a new Randstad will be built, and a new coastline with new large ports and new cities will be created.

Plan B is in fact not a real plan, for Pleijster. "It is an agreement to explore other options for the future beyond dikes, dams and polders. This Plan B projects a new Netherlands without dikes and polders. The west coast protects a the central marine lagoon. In true tradition of the Dutch, this lagoon is used for housing, fishing, recreation. By literally living with water, and building with nature, the Dutch might even come out stronger."

材料和风格组成一个色彩缤纷的混合体,充满着无限的活力,而且意外地在社交媒体上引人注目。

或许金融危机持续的时间不够长,尚不足以让这类创新举措从缓慢的发展中获益;而经济复苏后,地价和房市的压力又急剧增加,导致荷兰大城市的房地产市场几乎来不及喘息。但无论如何,房地产开发者都应抱有更多的耐心,这样才能创造价值。

"所有投资者都应该看清,接纳自然的建筑尽管从短期来看成本较高,但最终会带来收益。"鹿特丹LOLA景观建筑师事务所三位创始人之一的埃里克·简·普莱斯特说道,"我们应该把城市当作自然保护区来看待。而在当下,鹿特丹这样的城市就是全国最大的污染源之一。"

LOLA开发了一个将自然融入城市的新工具。城市及其建筑如同多个生态区,每个生态区都可以通过给本地昆虫、鸟类、蝙蝠等物种增加筑巢的可能性,来得到整体的提升。但最重要的是,要让植物在建筑内外蔓延。普莱斯特表示,"这样一来,城市中的自然体验就得到了放大,人们也能快乐、健康地生活。这最终也会体现在房产的价值上。"

这个工具是普莱斯特所说的A计划的策略之一。但当气候变化的副作用并没有因《巴黎协定》得到缓解时,就需要制定B计划。"在关于2200年的B计划中,荷兰将顺应自然。如果海平面急剧上升,我们就不得不把大部分土地还给海洋和河流三角洲。或许我们可以在西部保留一些历史悠久的城市中心",普莱斯特说。但东面就必须建设一个新的兰德斯塔德,创造一条分布有大型港口和城市的全新海岸线。

事实上,对普莱斯特来说,"B计划"并不是一个真正的计划,"它是对堤坝和围垦地之外的更多可能的探索。B计划设想了一个没有堤坝和围垦地的新荷兰。西海岸保护着中部的海岸潟湖,而在荷兰的传统中,这里本就用于居住、捕鱼和休闲娱乐。通过真正的与水共生,与自然共建,荷兰人甚至可能变得更加强大。"

Bob Witman was chief-editor of the art department of the Dutch daily paper *De Volkskrant*. Writes on architecture since 1999. (co-) Published books on public space *Amsterdamse Pleinen*, Vinexwijken *Vathorst Amersfoort*, the history of Dutch public buildings *De Werken, 90 Jaar Rijksgebouwendienst* and has been engaged with What Design Can Do on the social impact of design.

鲍勃·惠特曼曾任荷兰日报《人民报》的艺术主编,自1999年起从事建筑方面的写作。(合作)出版了关于公共空间(《Amsterdamse Pleinen》)、VINEX地区(《Vathorst Amersfoort》)、荷兰公共建筑史(《De Werken, 90 Jaar Rijksgebouwendienst》)等方面的书籍,参与了"What Design Can Do"项目,关注设计的社会影响。

bureau SLA, ZakenMaker
Co-living Oosterwold
Almere 2014–2017

布洛SLA，扎克马克
奥斯特沃尔德共享住宅
阿尔默勒 2014-2017

Bureau SLA and ZakenMaker designed 9 homes for a group of pioneers wishing to turn a potato field into a community.

In the rural area of Oosterwold Almere, artist Frode Bolhuis dreamed of an alternative way to live and work. He wanted to make a one-hectare potato field into his home and asked bureau SLA and ZakenMaker to design his dream house within a very limited budget. The architects came up with 2 preconditions to make the project possible. First, they suggested to Frode to find some friends to join the project. It is a lot cheaper to build several houses at the same time than just building one. The second precondition was that only the exterior would be designed, leaving the families complete freedom to decide on the interior.

The end result is a convincing piece of architecture. The 100-meter-long building makes a statement in the landscape. At the same time, it accommodates 9 completely different homes. All the families were allocated 160 m² of living space. This allowed them to have their own needs met – for example, by incorporating an artist's studio, or a large living room. The architects also decided to raise the long building off the ground so that it seems to float above the land. This design decision had another advantage. It allowed the residents to choose where their sewage system and water pipes would be located.

To achieve high-end insulation within the budget, the architects chose to get the most out of common building materials. Floor, roof, and adjoining walls are built as hollow wooden cassettes, and insulating cellulose was pumped in only upon completion. The result is an exceptionally well-insulated building.

The facade is designed as a strategy to give maximum freedom of choice within an efficient building system. Each family received a catalog of 7 windows and doors which could be placed anywhere in the facade. The space between the frames is vitrified with solid parts of glass without a frame. This creates an uncluttered but diverse facade.

The position of the building on the side of the plot leaves maximum space for a large community garden, and the long, communal porch makes it easy to make contact with the neighbors.

Oosterwold Co-living Complex demonstrates that it is possible to achieve a convincing design within a tight budget which, most importantly, manages to meet the expectations of 9 different clients at the same time. In the end, what had first seemed like a problem – the lack of building budget for interiors – became a key feature of this project. Completing the interiors of the homes strengthened the community. After all the hard work and its completion in summer, in the lee of the surrounding forest edge, one meter above ground level, the residents look over to enjoy their shared landscape and vegetable garden.

Generic plot 通用绘图

18% Building 建筑
8% Infrastructure 基础设施
13% Public green 公共绿化
2% Water 水
59% Agriculture 农业

1. Building 建筑 18%
   (storage 仓库)
2. Infrastructure 基础设施 8%
3. Public green 公共绿化 13%
4. Water 水 2%
5. Agriculture 农业 59%
   (helophyte filter 过滤沼生植物)

Credits and Data
Project title: Co-living Oosterwold
Client: Stichting Bosveld (Frode Bolhuis)
Location: Emile Durkheimweg, Almere, the Netherlands
Design: 2014–2015 (first project), 2017 (second project)
Completion: 2017 (both projects)
Architect: (bureau SLA) Peter van Assche, Ninja Zurheide, (ZakenMaker) Mathijs Cremers
Project team: Frode Bolhuis (project management), Bouwbedrijf Siemensma (contractor), Van Zuilen Constructie Advies, W2N engineers (engineering), Van der Weele Advies (sustainability advice)
Project area: 1,450 m² (gross floor area)
Project estimate: 700,000 euros (without tax)

pp. 38–39: View from the southwest. A 100 meter long collective housing designed to fit 9 families. pp. 40–41: Close up view of the dwelling unit. Photo by Toon Mijke. Opposite, above: Generic design based on the land use requirements of this district. Opposite, below: Diagram of the final proposed design by bureau SLA and ZakenMaker which also adhere to the same land use requirements. This page, above: View from the southeast. In addition to the shared farmland and warehouses, a pond with a water filtration system is placed in front of the building. Photos on pp. 38–47 by Filip-Dujardin unless otherwise specified. This page, below: Oosterwold district master plan by MVRDV. The D.I.Y approach of the urban plan allows residents to use their land freely. Image courtesy of MVRDV.

第 38-39 页：西南方向外观。100 米长的集体住宅容纳了 9 个家庭。第 40-41 页：居住单元近景特写。对页，上：基于该地区土地使用要求的通用设计；对页，下：布洛 SLA 和扎克马克在遵循同样的土地使用要求的情况下，提出的最终方案设计图。本页，上：东南方向外观。除了共享的农田和仓库外，建筑前方还设有一个带水过滤系统的池塘；本页，下：出自 MVRDV 建筑师事务所的奥斯特沃尔德地区总体规划。城市规划的 D.I.Y 策略允许居民自由使用他们的土地。

布洛 SLA 和扎克马克为一群希望把马铃薯田打造成社区的拓荒者们设计了 9 所房子。

在奥斯特沃尔德阿尔默勒的乡下，艺术家弗罗德·布尔威斯渴望着另一种生活和工作方式——他想将一片占地 1 公顷的马铃薯田纳入自己家中。他请来布洛 SLA 和扎克马克，在预算十分有限的条件下帮他设计出梦想中的房子。为了使项目的实现成为可能，建筑师提出了两个先决条件：首先，他们建议弗罗德找一些朋友加入这个项目，因为同时建造多栋房屋要比单独建造一栋房屋单价便宜得多。第二个前提条件是只设计外观，室内设计由住户全权决定。

最终成果是一座令人叹服的建筑。这座 100 米长的建筑从景观中脱颖而出。与此同时，它容纳了 9 个完全不同的家庭，每个家庭都拥有 160 平方米的居住空间。这使得他们各自的需求都能得到满足，比如在家中划出一间艺术家工作室，或规划一个超大的起居空间。建筑师还将这座长条建筑物抬离地面，使其看起来仿佛漂浮于地面之上。这样的设计决策还有另一个优势，即使住户可以选择设置污水处理系统和水管道的位置。

为了在预算范围内实现高品质的隔热效果，建筑师选择了充分利用普通建筑材料的优势。地面、屋顶和毗连的墙壁被建成了中空的木盒子，隔热纤维素在完工前才被泵入。最终得到了一座隔热性能绝佳的建筑。

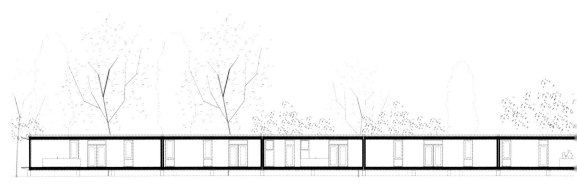

Long section (scale: 1/400)／纵向剖面图（比例：1/400）

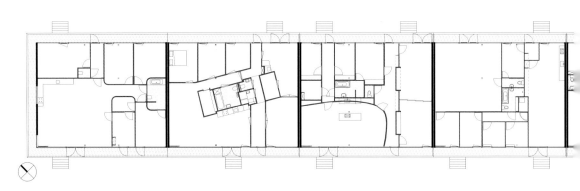

Plan (scale: 1/400)／平面图（比例：1/400）

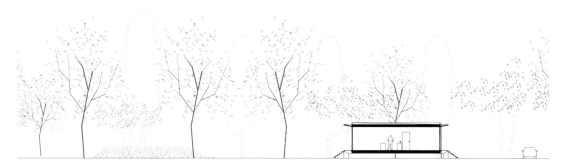

Short section／横向剖面图

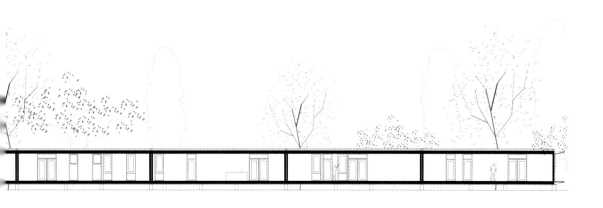

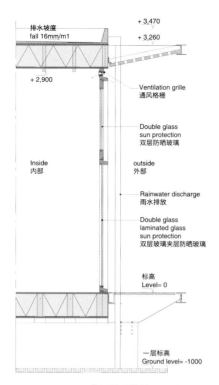

Section detail (scale: 1/50)／剖面详图（比例：1/50）

立面的设计策略在高效的建筑系统中给予住户最大程度的选择自由。每个家庭都被给到一份包含7种不同门窗样式的产品目录，他们可以从其中任选，将其置入立面上的任意位置。框架之间的空隙用牢固的无框玻璃进行封闭。如此，创造出一个整洁且多变的立面。

建筑位于地块的一侧，为大型社区菜园留出了最大的空间；此外，长长的共享门廊使邻居之间的交流变得更加容易。

奥斯特沃尔德共享生活集合建筑展示了在预算紧张的情况下，依然能够完成令人折服的设计。最重要的是，它同时满足了9个不同客户家庭的期望。起初看起来很成问题的问题——室内设计预算不足，最终成了这个项目的重要特征。一起完成的房屋内部装修促进了社区的融合。经过所有艰苦的工作，室内装修终于在夏季竣工；在周围森林边缘、高出地面1米的庇荫处，住户们欣赏着他们共享的景观和菜园。

Opposite, both images: Interior views of the dwelling unit. The interior can be freely laid out by residents. Each unit has a floor space of 160 m². Photo by Linda Merlijn. This page: View from the northwest end of the building. The floor level is raised 1 m off the ground with an equipment space provided underneath.

对页，两图：住宅单元内部。室内可由住户自由布置。每个单元的占地面积均为160平方米。本页：从建筑的西北端看建筑。地板被抬高了1米，下方设有设备间。

Space&Matter
De Ceuvel
Amsterdam 2012–2014

空间&变体建筑事务所
德·苏维尔
阿姆斯特丹 2012–2014

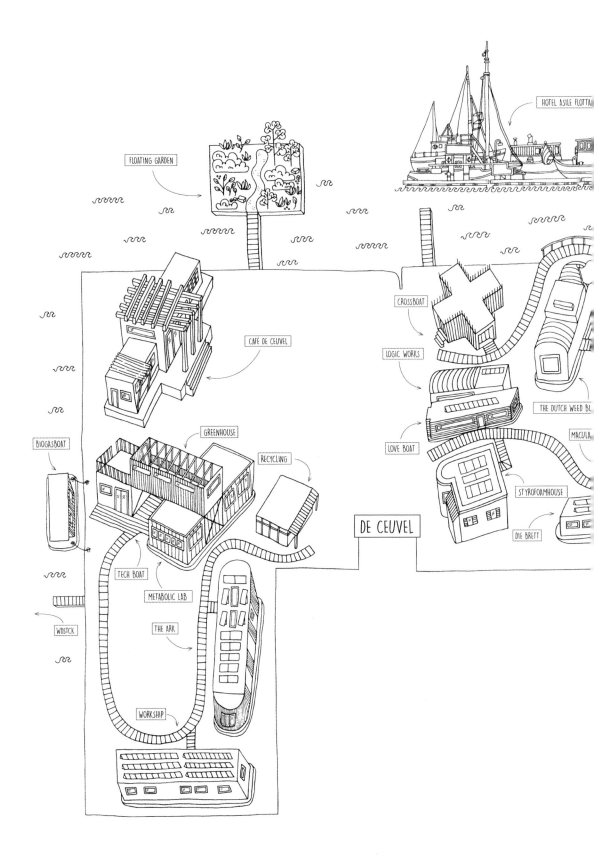

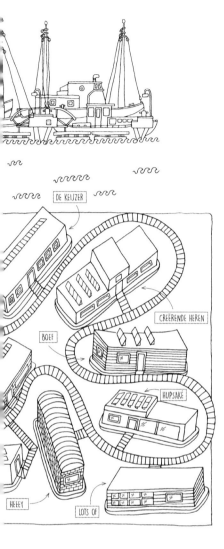

In 2012, the municipality launched a tender for the regeneration of a former shipyard that turned into a post-industrial wasteland in North Amsterdam. The conditions of the tender were simple: It had to provide a workplace for creative industries, and have an underlying focus on sustainability. However, the conditions of the site were less straightforward. The soil quality was so poor that digging into the ground was prohibited, making it impossible to apply conventional building foundations and sewage and water systems.

Space&Matter and partners won the tender with an ambitious plan centered around 16 retrofitted houseboats. The houseboats are connected by a raised boardwalk which acts as a communal space for the offices. It also conceals technical components of the site which would otherwise be underground. Underlying the development is a garden of soil-cleansing plants that, through a process of phytoremediation, naturally extract the harmful metals from the ground.

Referred to as a "cleantech playground", De Ceuvel is a space for experimentation. With low tech innovations such as composting toilets, self-made heat pumps, and water filtering systems, it explores the possibilities of off-grid living and decentralized systems. It also experiments with energy sharing through the use of solar panels connected to a smart grid, which allows users to trade energy locally.

With its learning-by-doing approach, De Ceuvel acts as a community test lab to explore innovative approaches to sustainability and uses these approaches to inspire and mobilize others. It is celebrated nationally and internationally as a frontrunner of circular living.

pp. 48–49: View from the boardwalk that connects the boathouses. Boathouses are used in creating the co-working spaces for a creative industry as foundation work cannot be done due to soil contamination. Photo courtesy of Martijn van Wijk. Opposite and this page: Illustrated map of De Ceuvel. Images courtesy of De Ceuvel.

第48-49页：连接船屋的木板路外观。由于土壤污染，无法实现基础工程，因此，这些船屋被用作创意产业的协同工作空间。对页及本页：德苏维尔的地图导示。

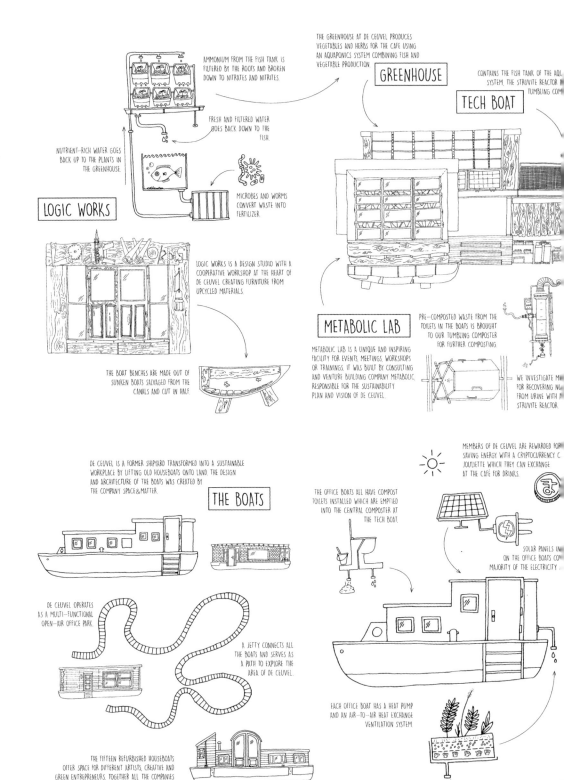

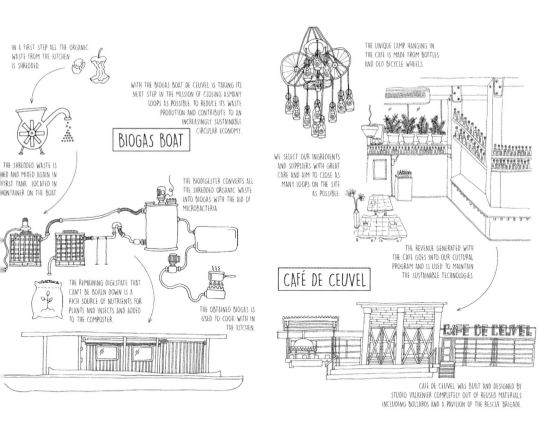
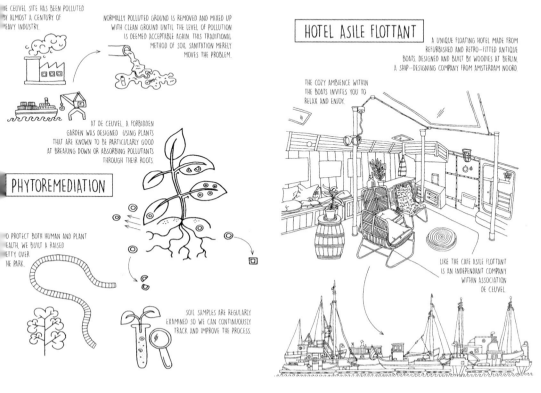

pp. 52-53: Illustration showing the functions of each boat. Images courtesy of De Ceuvel. This page, from above to below: A scene of De Ceuvel during an event. A live show was held at the cafe. Both photos courtesy of De Ceuvel. View from the entrance in the northwest, where the cafe and a front open space can be seen. Photo by Martijn van Wijk. Opposite: Aerial view from the south. Photo courtesy of De Ceuvel. p. 56, above: Photo taken during the transportation of the boathouse. Photo by Jean-Pierre Jans. p. 56, below: Construction work on the boathouse's interior. p. 57: Construction of the broadwalk. Technical components are stored underneath the boardwalk. Both photos by Martijn van Wijk.

第 52-53 页：每间船屋的功能示意图。本页，从上至下：活动期间德·苏维尔的场景。咖啡馆举办现场表演。从西北入口处可以看到咖啡馆及其前部的开放空间。对页：南面鸟瞰图。第 56 页，上：船屋的运输过程留影；第 56 页，下：船屋的内部建设。第 57 页：木板路的建造。技术构件藏于木板路下方。

2012年，市政府对阿姆斯特丹北部一处老旧造船厂进行的条件很简单：它必须为创意产业提供工作场所，并将重点放在可持续性上。然而，这块场地的条件复杂。由于土壤质量太差，向地下挖掘受到限制，因此无法使用传统的建筑地基和污水、供水系统。

空间&变体建筑事务所及其合作单位，凭借一个以16艘改造船屋为中心的宏大设计方案中标。船屋通过高于地面的木板路相连，而这些连接也为办公室提供了公共空间。它还隐藏了现场那些原本应该埋于地下的技术构件。项目开发的前提是建造一个净化土壤的植物园，通过植物修复过程，地下的有害金属被自然地吸收。

德·苏维尔被称为"清洁技术乐园"，也是一个实验场所。通过诸如堆肥厕所、自制热泵和滤水系统等低技术创新，它探索了离网生活和分散式系统的可能性。它还使用连接智能电网的太阳能电池板进行能源共享试验，使用户可以进行小范围的能源交易。

通过实践学习的方式，德·苏维尔充当着社区的测试实验室，探索着实现可持续的新方法，并利用这些方法来激发和动员其他人。它在本土和国际上都被誉为"循环生活的领跑者"。

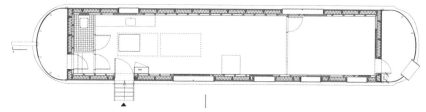

Floor plan／平面图

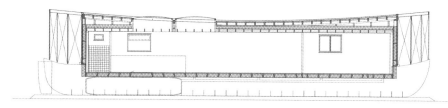

Long section／纵向剖面图

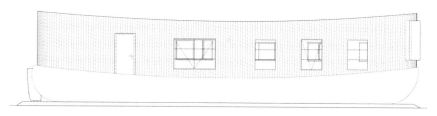

Long elevation (scale: 1/200)／纵向立面图（比例：1/200）

Credits and Data
Project title: De Ceuvel
Client: Self-initiated by Space&Matter and partners
Location: Amsterdam, the Netherlands
Design: 2012
Completion: 2014
Architect: Space&Matter
Design team: Sascha Glasl, Tjeerd Haccou, Marthijn Pool, Federica Heimler, Martijn van Wijk, Rogier van den Brink
Project team: Space&Matter (urban plan and design, architecture, project development), Smeelearchitecture (project development, community), Jeroen Apers architect (project development, finances), Marcel van Wees (project and general management), Metabolic (concept development, research, and implementation of clean technologies and sustainability plan), DELVA Landscape Architects & Bureau Fonkel (design, implementation, and maintenance of the Purifying Park), Studio Valkenier (design of Café de Ceuvel), Waterloft (financial advice)
Project estimate: 450 euros (with sweat equity)

Hilberinkbosch Architects
The Sixteen-Oak Barn
Berlicum 2017–2018

希尔伯林克·博世建筑师事务所
十六号橡树谷仓
拉里克姆 2017–2018

Haphazard aesthetics

In 2017, we were told that 7 of the century-old oak trees in our yard were in bad shape and had to be cut down. Instead of following the usual path of selling the trees to the paper industry, we decided to reinstate an ancient tradition. And so, by replacing a collage of obsolete shelters and sheds, we would – in line with our farm's monumental character – build a new barn with locally harvested materials employing traditional techniques.

Our new transverse barn is constructed with 4 tie-bar couple trusses that are connected by longitudinal stringers, topped with a roof supported by rafters. The roof has an asymmetrical geometry with a steep and a low pitched side. There are 3 spaces inside the barn: a carport, a storage room, and a workshop/meeting room for office use. The loft above the storage room opens to the workshop.

Untreated timber, concrete, and glass have been intermingled in various ways. The irregular dimensions of the wood used to build the formwork resulted in far from perfect concrete surfaces. Remnants of the sapwood formwork became embedded in the concrete and tannic acid from the fresh timber left discolorations. Iron and steel inclusions in the trees such as barbed wire and shrapnel, presumably from the World War in 1944, brought imperfections to the timber. These additional circumstances created an unexpected layer of traces in the wood and concrete.

The barn's aesthetics have been strongly influenced by coincidence. It lends this contemporary building a vital expression that merges the old and new in a wonderful and extraordinary way.

Credits and Data
Project title: The Sixteen-Oak Barn
Location: Berlicum, the Netherlands
Design: 2016
Completion: 2018
Architect: Hilberinkbosch Architects
Design team: Annemariken Hilberink, Geert Bosch, Frenske Wijnen, Jaap Janssen
Project team: Zandenbouw, Aarle-Rixtel (contractor), Brabanthout, Den Dungen (mobile sawmill), Raadgevend ingenieursburo van Nunen (structural engineer), Parklaan Landschapsarchitecten (landscape)
Project area: 123 m²

1. Columns／柱

2. Stringers／桁

3. Slats／条板

4. Braces／角撑

5. Stair steps／踏板

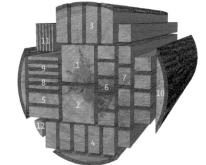

6. Rafters／椽

7. Glazing bars／玻璃隔条

8. Roofing boards／望板

9. Cladding／外墙板

10. Formwork／模板工程

11. Firewood／柴薪

12. Shingles／木瓦

偶然美学

2017 年，我们得知院子里的 7 棵百年古橡树由于状况不佳，不得不被砍掉。而后，我们并没有依照常规把树木出售给造纸商，而是决定利用它们恢复这里古老的传统。因此，我们就地取材，利用传统工艺，秉持农场的重要特征，建起了一座新的谷仓来代替这里已经废弃的庇护所和棚屋。

新建的横向谷仓由四根拉杆连接的桁架构成，这些桁架通过纵向桁条连接，其上是由椽子支撑的屋顶。屋顶是不对称的几何结构，坡屋面一侧较陡，另一侧则较平缓。谷仓内容纳了 3 种空间，即车库、储藏室和供办公使用的工作间和会议室，储藏室上方的阁楼空间向工作间敞开。

未经处理的木材、混凝土和玻璃以各种方式交织在一起。用于建造浇筑模板的木材尺寸各异，导致混凝土表面不够完美。模板边材的残留物被嵌入到混凝土中；除此之外，新鲜木材中含有的单宁酸会导致木材褪色并沾染在混凝土表面。树木中的钢铁夹杂物，例如铁丝网围栏和那些可能是 1944 年"二战"时期留下的炮弹碎片，使木材有了瑕疵。正是这些额外的素材条件，在木材和混凝土上留下了意想不到的痕迹。

谷仓的美受到了一系列偶然的强烈影响。这些偶然为这座当代建筑增添了一种至关重要的表达方式，并以奇妙又非凡的方式使新旧得以融合。

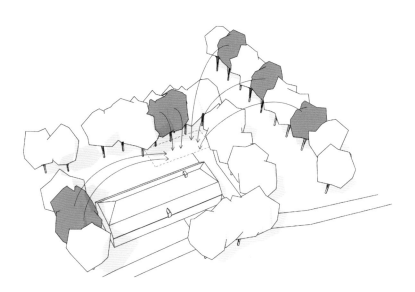

Site diagram. The barn is made out of the old oak trees in the surrounding yard.
基地分析图。谷仓是用周围院子里的老橡树建造起来的。

p. 59: Southwest facade. Remnants of the sapwood formwork became embedded in the concrete leaving behind discolorations. Opposite: Diagram illustrating the use of each timber. Photos on pp. 58–67 by René de Wit.

第 59 页：西南立面。模板边材的残留物被嵌在混凝土中，留下了变色痕迹。对页：每一根木材的使用图解。

1. Workshop - meeting room	1. 工作间 - 会议室
2. Storage	2. 储藏室
3. Carport	3. 车库
4. Concrete worktop	4. 混凝土台面
5. Woodstove de Klein & van Hoff	5. 德克莱恩和范霍夫的壁炉
6. Firewood in recess	6. 存放柴薪的凹槽
7. Loft floor	7. 阁楼

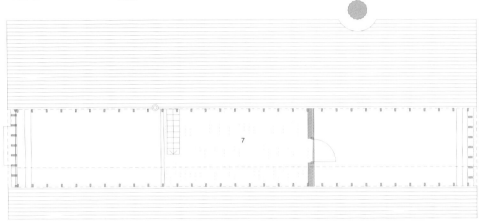

Upper floor plan／上层平面图

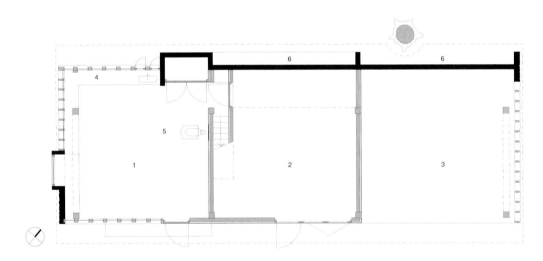

Ground floor plan (scale: 1/200)／一层平面图（比例：1/200）

Opposite, above: Southeast facade. Timber cladding is used on this side of the facade. Opposite, below: Northeast facade.

对页，上：东南立面。整个立面均由木材覆盖；对页，下：东北立面。

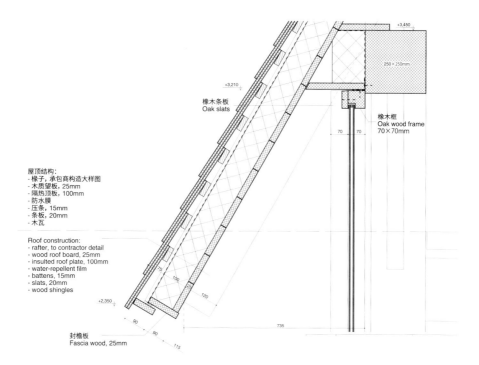

Roof detail (scale: 1/15)／屋顶细部详图（比例：1/15）

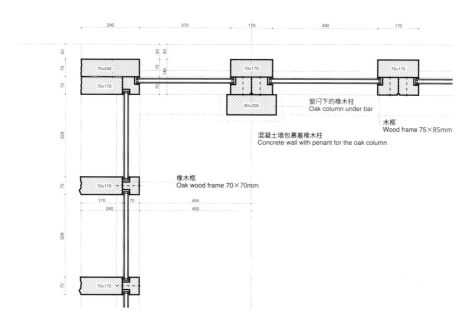

Window detail (scale: 1/15)／窗户细部详图（比例：1/15）

pp. 64–65: *Interior view of the workshop and meeting room. Opposite, above: View of the storage and a loft space above. Opposite, below: Close up view of the concrete worktop and oak cladding.*

第 64-65 页：工作间和会议室内部。对页，上：储藏室及其上的阁楼空间外观；对页，下：混凝土台面和橡木外表皮近景。

RAU architects, RO&AD architects
Tij Observatory
Stellendam 2019

RAU建筑师事务所，RO&AD建筑师事务所
Tij鸟类观测站
斯泰伦丹 2019

pp. 68-69: Aerial view of the observatory in Scheelhoek. Photo courtesy of the architects. This page: The form of the observatory is modeled after a sandwich tern egg.

第 68-69 页：希尔斯霍克自然保护区内的鸟类观测站鸟瞰图。本页：蛋形的鸟类观测站是参照白嘴端凤头燕鸥的鸟蛋形状建造的。

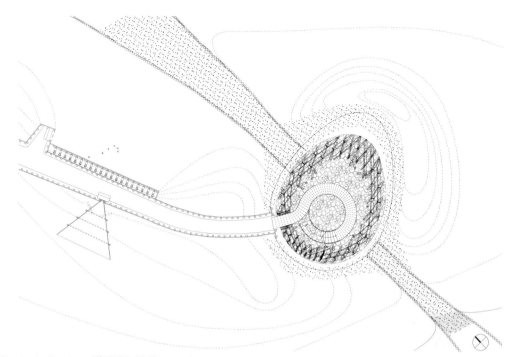

Site plan (scale: 1/300)／总平面图（比例：1/300）

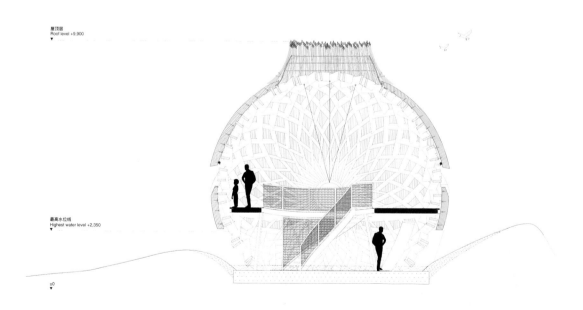

Observatory section (scale: 1/150)／观测站剖面图（比例：1/150）

71

Tij is an object designed to celebrate the opening of the Haringvliet sluices in November 2018. The sluices were opened in order to improve water quality and biodiversity while stimulating fish migration from the North Sea to the river delta system of Maas and Rhine in the Netherlands. To provide people with the opportunity to experience and explore these changes, a bird observatory is built in the Haringvliet area.

Tij is an egg-shaped bird hide situated in Scheelhoek, a nature reserve close to the Haringvliet sluice near Stellendam. The islands of Scheelhoek are breeding and feeding grounds for several species of birds like the common tern, spoonbill, and the icon of this area, the sandwich tern.

To prevent the birds from being disturbed, the route to Tij is hidden and the last section of the path is, in fact, a tunnel made of re-used mooring posts and second-hand azobe planks which were once used in the brick industry. The tunnel is covered in sand to provide habitat for the terns or waders. The exterior of the tunnel provides artificial nesting holes for sand martins. The endpoint of the route is in Tij, where you can view hatching terns and other birds.

The egg is modeled on a sandwich tern egg, and sits in a nest, much like what a tern would have done. The nest of the egg consists of vertical "feathers" of chestnut poles, reeds, and small sand dunes. Tij is parametrically designed to achieve a good ratio between form, structural integrity, size of the timber, and the size of the openings. The structure has been constructed as a File-to-Factory Zollinger to provide relatively big spans with small timber parts. It can be completely disassembled. Through its re-usability, its modularity, its materials, and its contribution to the natural environment, it is almost completely circular and sustainable.

Credits and Data
Project title: Tij Observatory
Client: Vogelbescherming & Natuurmonumenten
Location: natuurgebied De Scheelhoek, Stellendam, the Netherlands
Design: October 2018
Completion: March 2019
Architect: RAU Architects (Main Architect), RO&AD Architects,
Design team: (RAU) Thomas Rau, Michel Tombal, Jochem Alferink, (RO&AD) Ad Kil, Ro Koster, Martin van Overveld, Athina Andreadou, Loyse Rebord, Rodrigo Altamirano
Project team: BreedID (main structural engineer), Aalto University Finland (structural engineer wood), Geometria (wood engineering), H+N+S (landscape), Van Hese Infra (contractor), Elg Rietdekkers (thatched roof)

Opposite, both images: Interior of the observatory. Photos by Katja Effting.

对页，两图：观测站内部。

This page, both images: Photos taken during the construction of the File-to-Factory Zollinger structure. Photos courtesy of the architects.

本页，两图：工厂预制的结构构件在现场进行拼装的施工照片。

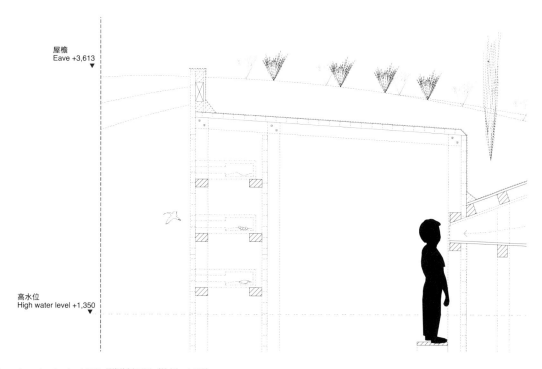

Tunnel section (scale: 1/50)／隧道剖面图（比例：1/50）

Tij（荷兰语"浪潮"之意）鸟类观测站是为庆祝2018年11月哈灵水道水闸启用而建的。水闸的开放旨在改善水质并增强生物多样性，同时促使鱼类从荷兰北海迁徙至马斯和莱茵河的三角洲地带。为了鼓励人们体验并探索这些变化，哈灵水道地区新建了这座鸟类观测站。

Tij是一个蛋形的鸟类栖息之所，位于哈灵水闸附近的希尔斯霍克自然保护区，靠近斯泰伦丹。这里的岛屿是许多鸟类的繁殖地和觅食区，比如，普通燕鸥和篦鹭，以及当地标志性的白嘴端凤头燕鸥等。

为了避免对鸟类造成干扰，通往Tij的游览路线很隐蔽，事实上，该路线的最后一段甚至被设计成藏于地下的隧道。隧道由回收再利用的系船柱和曾用于制砖业的红铁木板建造而成。隧道内表面被沙子覆盖，能够为燕鸥或涉禽提供栖息环境，外表面则为崖沙燕提供了筑巢的凹洞。这座鸟类观测站位于路线的终点，从这里可以观察到正在孵化的燕鸥和其他鸟类。

坐落在沙巢上的蛋形空间是参照白嘴端凤头燕鸥的鸟蛋形状建造的，如同天然的鸟巢一般。"鸟蛋"所在的巢由垂直排列的"羽毛"——栗木条和芦苇，以及小沙丘构成，"鸟蛋"本身则通过参数化设计实现了形状、结构完整性、木材尺寸以及开口尺寸之间的良好比例。结构构件在工厂预制完成后在现场进行拼装，以便通过小尺寸的木制组件来实现较大的跨度。这样的木结构可以完全被拆解。可重复利用的模块化属性、环保材料的使用和对自然环境作出的贡献，使它具备了近乎完美的循环性和可持续性。

RAAAF, Atelier de Lyon
Deltawerk//
Marknesse 2014–2018

RAAAF，里昂工作室
三角洲之作//
马克内瑟 2014–2018

A giant wave basin served as a test site for the Dutch delta works. Deltawerk// questions the ambition to build an indestructible Holland in times of climate change. In relation to this, the artwork is also an experiment in making new ruins.

The enormous test models in national monument Waterloopbos, the former Dutch Hydrodynamics laboratory, are in decay. One of these models is the Delta Flume; the waves in the Delta flume are gone after 40 years of technological progress and many different experiments. By excavating the sand plateau around the flume, a gigantic "delta work" of 7 m high and 250 m long is unveiled and surrounded by water. Massive concrete slabs are cut out of the 80 cm thick walls and turned 90 degrees around their axis. The space offers an intense spatial experience of light, shadows, reflection, and views through the Waterloopbos. The space changes throughout the day, the seasons, and over the years.

In a radical way, this intervention sheds new light on the Dutch and UNESCO policy on cultural heritage: Hardcore Heritage.

Through deliberate destruction, radical changes in context, and seemingly contradictory additions, a new field of tension arises between the present, past and future that activates built heritage, instead of "extracting" it from history and putting it on a pedestal. Such hardcore heritage interventions open up ways of interpreting history toward the future, rather than being stuck in fixated narratives from the past.

Credits and Data
Project title: Deltawerk//
Client: Land Art Flevoland & Natuurmonumenten Dutch Cultural Heritage Agency
Location: Waterloopbos National Monument, Marknesse, the Netherlands
Design: 2014
Completion: 2018
Architect / Artists: RAAAF and Atelier de Lyon
Design team: Erick de Lyon, Ronald & Erik Rietveld, David Habets, (support team) Cecile-Diama Samb, Johnny Long, Marina Fernandez, Joop Schroen
Project team: RAAAF, Atelier de Lyon (contractor), Dikkerboom (subcontractor),
Material: cut reinforced concrete and water
Project area: 18,000 m²

pp. 76–77: View from the northwest. An artwork is created by cutting out the walls at the Dutch Delta Works research facility. Opposite, above: View from inside the artwork, Deltawerk//. Opposite, below: View from the area between the closed walls on the south side. Photos on pp. 56–61 by Jan Kempenaers.

第 76-77 页：西北面外观。该艺术装置利用从荷兰三角洲工程研究设施的墙壁上切取的部分制成。对页，上："三角洲之作//"的内部；对页，下：南侧封闭墙壁之间的区域。

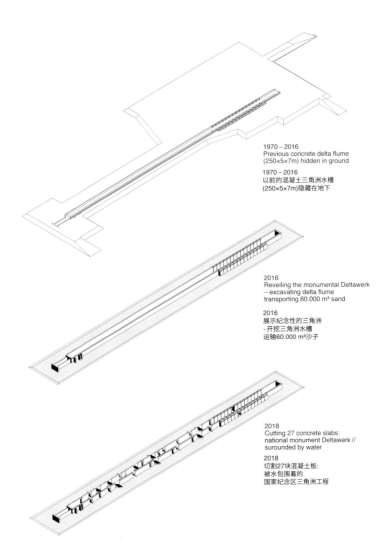

1970 – 2016
Previous concrete delta flume
(250×5×7m) hidden in ground

1970 – 2016
以前的混凝土三角洲水槽
(250×5×7m)隐藏在地下

2016
Reveiling the monumental Deltawerk
– excavating delta flume
transporting 60.000 m³ sand

2016
展示纪念性的三角洲
-开挖三角洲水槽
运输60.000 m³沙子

2018
Cutting 27 concrete slabs:
national monument Deltawerk //
surounded by water

2018
切割27块混凝土板：
被水包围着的
国家纪念区三角洲工程

地基内的巨大水路曾经是荷兰三角洲工程的试验场。"三角洲之作 //"质疑着在气候变化时期，建设一个坚不可摧的荷兰的雄心壮志。同时，该艺术作品也是对创造新古迹的一次尝试。

存在于国家纪念区水道森林内，前荷兰流体动力学实验室中那些巨大的测试模型已经衰败。其中一个模型是三角洲水槽。历经 40 年的技术改进和大量不同的实验，三角洲水槽中的水波早已不复存在。通过对水槽周围沙土高原的挖掘，一个 7 米高、250 米长的、被水环绕的巨大"三角洲"出现在公众视野中。巨大的混凝土板是从 80 厘米厚的墙中切下来的，并绕其轴线旋转了 90°。

这里有着光影交错的强烈空间体验以及遍布水道森林的景观。一天之中、一季之中、一年之中该空间不断发生着变化。

从根本上讲，这样的介入为荷兰和联合国教科文组织的文化遗产政策"核心遗产"，提供了新的思路。

设计师通过有意的破坏、对环境的彻底改变以及看似矛盾的添加，在现在、过去和未来之间制造出一种全新的张力，以此来激活建筑遗址，而不是将其从历史中"提取"出来并置于基座之上。这种对核心遗产的应对方式开辟了一种面向未来诠释历史的新途径，而不是一味陷入对过去的讲述中。

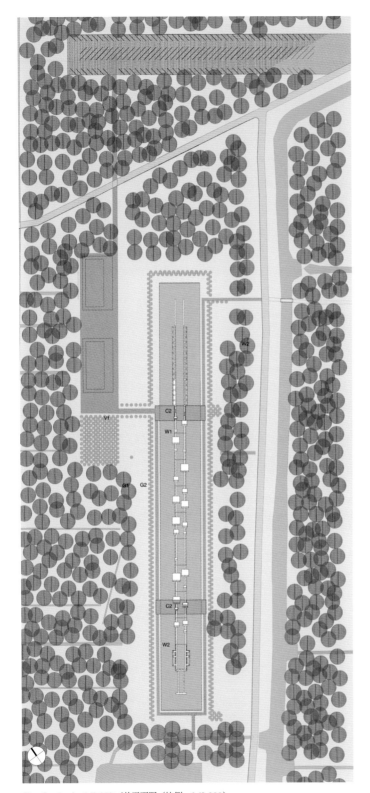

- V1. Rubble path / 碎石小道
- C2. Solid concrete / 实体混凝土
- W1. Water / 水
- W2. New transit water tube / 新运输水管
- G2. New indicative trees / 新的指示树

Site plan (scale: 1/2,000) ／总平面图（比例：1/2,000）

p. 81, above: Photo of the giant wave basin in the research facility before the artwork intervention. p. 81, below: Aerial view of the site taken during 2016. Excavating the surrounding soil revealed its 7 m high walls. This page: Aerial view from the west. The artwork is located in Waterloopbos nature park.

第 81 页，上：在该艺术作品介入之前，研究设施中巨大水路的外观；第 81 页，下：拍摄于 2016 年的场地鸟瞰图。挖掘周围的土壤，从而使 7 米的高墙显露出来。本页：西面鸟瞰图。该艺术作品位于水道森林自然公园。

Essay:
Tabula Scripta
Jarrik Ouburg

"Don't avoid the complexity of reality, let it take you beyond your personal and professional comfort zone."
Anonymous student Workshop GUC Academy of Architecture Amsterdam, Cairo, 2017

Tabula Scripta is a study by Floris Alkemade, Michiel van Iersel and Jarrik Ouburg at the Academy of Architecture Amsterdam. This study – spanning several years – sees the complex reality in which we live and work as an inscribed page that we must read and understand in order to write more on it. The study's subject is the attitude and working method of the architect who does not look at the layered context of a project as a parameter that lies outside the design brief, but as that which defines the design brief itself.

A Bad Start

From an economic perspective, 2012 was the worst year for Dutch architects in decades. Architectural firms lost half of their business and many architects found themselves out of work.[1] The crisis which the once so illustrious Dutch architecture found itself in was remarkable and, in my view, had a twofold cause: it was a "double dip".

Dutch architects had become very dependent on commercial commissions and, following the 2008 financial crisis, their clients had reached the end of their financial reserves by 2012. There were hardly any investments in building and therefore architects were getting much fewer commissions. It became painfully clear that, through the years, building had increasingly become an economic act and much less a cultural or societal effort. It had become a goal in itself, a tool for creating square metres and making money off them; it was no longer an instrument to raise the quality-of-life, provide decent housing, or create comfortable workplaces. The clearest illustration of this would be the new offices that had been built and left to stand empty. By making themselves dependent on a finite economic revenue model, Dutch architects had literally and figuratively priced themselves out of the market.

In addition to this cause, which could be argued to lie outside of the domain of architecture, there was a second cause that played a more intrinsic part in its domain. This was not because of an over accommodating attitude or modesty, but rather the opposite. During its heyday, the architectural profession had perhaps placed itself on too high a pedestal, as illustrated by the term "starchitect" – the architect as a pop star performing his trick. In the Netherlands, we took this one step further and elevated architecture itself to the same height. In 2000, the book SuperDutch was published which presented Dutch architecture as the super architecture. This designation was placed over our architectural climate like a glass dome, presenting it as an exhibit, but gradually this climate began to suffer from a lack of oxygen. More and more, designing was about the need to perform. The radical image of a building became more important than its societal or functional qualities. Architecture became a goal in itself and no longer a means to achieve that goal. Building for building's sake, and architecture for architecture's sake. They were two bubbles that burst at the same time.

Coincidentally, 2012 was also the year I

不要回避现实的复杂性,让它带你走出个人和职业的舒适区。"

GUC工作坊的某学生
阿姆斯特丹建筑学院,开罗,2017

Tabula Scripta(译注:意为"刻有文字的石板"或"有脚本")是弗洛里斯·阿尔克马德、米歇尔·凡·伊尔瑟和贾里克·奥伯格在阿姆斯特丹建筑学院的一项研究,已经持续了数年。该研究认为生活和工作的复杂现实是已被书写的一页,我们必须先阅读和理解,才能写下更多的内容。具体而言,这是研究特定群体建筑师的态度和工作方法——在这些建筑师眼中,项目的复杂文脉不是无关紧要的参考因素,正是它决定了如何进行概要性的设计。

坏的开端

从经济角度来看,2012年是几十年来荷兰建筑师最糟糕的一年。建筑公司损失了一半的业务,许多建筑师失业[1],一度辉煌的荷兰建筑业发现自己正处在显而易见的危机之中。而我认为,这有两个原因,也可以说是一场"双谷衰退"。

荷兰建筑师非常依赖商业委托,但到了2012年,那些经历过2008年金融危机的客户在财务储备上已然达到极限,鲜有投资进入建筑业,建筑师获得的委托也大幅减少。而令人痛心的是,这些年的建造趋向一种经济行为,与文化和社会的进取背道而驰。它成了一个目标,一个搭建平方米并从中赚钱的工具;它不再是为了提高生活质量、提供体面住房或创造舒适的工作场所。最明显的例子就是那些被闲置的新办公室。在有限的经济收入之下,荷兰建筑师把自己困在了有价无市的境地。

这可以说是建筑之外的原因,此外还有第二个建筑内部的、更本质的原因。建筑师不是过分迁就、谦虚,而是恰恰相反。正如"明星建筑师"这个词所展现的,在建筑行业如日中天的时候,或许建筑师自己都高估了这一职业的地位,将其与表演艺术的人气巨星画上等号。而在荷兰,建筑本身也被抬升到了同样的高度。2000年,《超级荷兰》出版,书中将荷兰建筑称为"超级建筑"。此后,这个头衔如同玻璃罩一般包裹着建筑,使建筑成为一种展品。但渐渐其内部开始缺氧,建筑设计越来越倾向于表现,激进的外观变得比社会性、功能性更加重要。建筑自身变成一个目标,而不再是实现目标的手段。"为建造而建造""为建筑而建筑",这两个泡沫在同一时刻走向了破灭。

巧合的是,2012年我应邀担任阿姆斯特丹建筑学院的建筑系主任,任期4年。我的任务是让年轻的学生对建筑学充满热情,帮助他们做好准备,有朝一日在建筑领域成就辉煌。我爽快地接下了这个任务,却没想到还有一连串的后续问题。比如,在这样一个时代,连我们的未来和这个学科的存在都受到质疑,又应为了什么目的、以什么样的方式培养学生?这些问题并不容易回答。

法国诗人、哲学家保罗·瓦勒里曾经写下"L'avenir -n'est plus ce qu'il était"(未来不再是过去的样子)。这种想法可能会引发沉思,但它也是能量的来源。在棘手的现实之外,教育提供了创造平行世界的机会和自由。仔细想来,这恰恰是对那句话的阐释,尤其当我们能够为新一代建筑师做到这一点的时候。

建筑学,如同爱吹牛的闵希豪森男爵[2],如何靠自己的"辫子"从泥潭中抽身?这场危机带来的最深刻的教训或许就是,建筑师不应为自己的"辫子"洋洋自得,只顾在白纸上画出惊艳四座的草图,只知道设计一栋栋孤立的建筑。为了融入新的社会角色,形成新的态度,建筑师更应该研究

was asked to become head of the Department of Architecture at the Academy for Architecture Amsterdam, for a 4-year term. My task would be to make young students enthusiastic about and prepare them for a glorious career in architecture. It didn't take me long to say "yes". The question did, however, give rise to many subsequent questions, such as "how" and "to what end" does one train students during such times when our future and the very reason for the field's existence are being questioned. These questions took a little bit longer to answer.

"L'avenir n'est plus ce qu'il était" ("The future is no longer what it used to be"), the French poet and philosopher Paul Valéry wrote. A thought that may evoke nostalgia or melancholy, but at the same time, it can be a source of energy. Education, with all its opportunities and freedom to create a parallel world alongside the intractable practice, proved – on second thought – to be just the place to explore this, especially if one is able to do so with a new generation of architects for whom their future will soon become today.

How can architecture – like the Baron Munchausen[2] – pull itself out of the mire by its own "ponytail"? The greatest lesson to be learned from the crisis was perhaps that architects, in order to pull off this feat, should be less concerned with their own "ponytail", with the brilliant sketch on a virgin white sheet of paper, and with the building as a solitary object. In order to grasp the new societal role and develop a new attitude, architects should rather study the mire, which in our situation refers to the layered built environment that after generations of building, demolishing, and renovating, produced the complex context where we live and work. Architects should not see this changed context as a threat, but rather as the inspiration, motivation, and the origin of a project. If they do, the study of architecture no longer focuses on what is on the virgin white sheet, but on the inscribed page – the state of the Tabula Scripta.

In order to take this inscribed page as a starting point to write more on it, or perhaps delete parts of it, we must first read and understand it. To do this, we have formulated research questions and established research methods in the form of verbs (action words), by which we as architects can respond better and smarter to what is already there. This leads to actions and reactions with an awareness that they should add value to what is already there instead of diminishing it.

These verbs are, in no particular order:

To eliminate – how can value be added by removing material?
Architecture is often seen as an act of adding (constructions and materials) to the world. The opposite – taking these out of a building or city – is often a forgotten instrument for creating spatial quality and forging necessary links, with just a fraction of energy. In the age of sustainability, when some regions of the country are facing declining populations and empty buildings, this form of creative destruction should be taken more seriously.

To continue – how can we regard existing and new buildings as an intermediate phase waiting to be developed further?

泥沼,在这里便是指几代人建造、拆除、改建后的多层次的建成环境,这也是我们生活、工作所处的复杂文脉。对于建筑师而言,这种发生变化的环境不应是一种威胁,而应是一个项目的灵感、动力和起源。如果真能如此,建筑学的研究焦点就不再是空白的纸张,而是已被书写的一页——即Tabula Scripta。

为了在Tabula Scripta写下更多的内容,或者进行删除修改,我们必须首先阅读它、理解它。为此,我们确定了研究问题,并通过动词(行动词)的形式建立了研究方法,这让我们能够作为建筑师更灵活地处理已有的事物。我们也在行动和反馈中具备了这样的意识:应该增加已有事物的价值,而不是加以削弱。

这些动词如下所示,排名不分先后:

消除——如何通过去除材料来增加价值?

建筑通常被视为一种为世界添砖加瓦的行为。反之,将建材从建筑或城市中取出,通常就成了一种遗忘的工具——只需一点点精力,就能创造空间质量,形成必要的关联。在可持续发展的时代,当国家某些地区面临人口减少,建筑空置的情况时,更应谨慎对待这种创造性的破坏。

继续——无论建筑新旧,如何将其视为还处在有待发展的过渡阶段?

和其他地方一样,在荷兰,许多建筑的寿命都比设计者更长。建筑物不断被改造,与它最初的设计渐行渐远。因此,作为建筑师,我们习惯在前人工作的基础上进行建造。既然心知肚明,就不必想着设计出最终的目标形态,或是创造出符合艺术想象的孤立的建筑。相反,我们可以致力于设计中间的、开放的建筑,并期待它们未来的更多变化。

模糊——如何把建筑从对外观的执着中解放出来,让不可见性创造价值?

当我们在历史文脉中考察建筑行为时,对现代建筑的保守态度似乎占了上风。一如1985年的电影《回到未来》里的那句话,因循守旧地建造新楼,似乎是建筑师们最安全的选择。然而,当把焦点从可见处转移到古老外墙背后隐藏的不可见处时,我们恰恰会发现,那里的空间在建筑、室内、流程和用途上都突破了常规。以古城阿姆斯特丹为例,正是没有历史负担的地下,提供了很多意想不到的实验空间,堪比现代大教堂的地下车站和自行车停车场,还赢得了建筑界最负盛名的奖项。

重构——如何架构我们的建成环境,将它视为有待重整的材料的临时集合?

新建筑的设计往往有着特定的目的和形式。在建造过程中,建筑元素(立柱、地板、墙体、螺钉)会被永久固定,以实现最终的结果,整体的价值也会超过各部分的总和。但如果建筑整体不再服务于它的初衷,其各部分的剩余价值将会如何呢?事实上,旧建筑材料的再利用有着悠久的传统,希腊和罗马建筑所谓的"spolia",便是指那些被拆解后又用于新建筑的部件。也许,我们需要探索一种新的建筑类型——在这种建筑中,部件及其再利用性将决定整个建筑的面貌。

再利用——如何通过赋予现有空间新用途,重新定义建筑?

建筑的定义有很多,其中之一便是形式和功能、空间

In the Netherlands, as elsewhere, many buildings outlive their designers. Buildings are constantly adapted, departing from its original design. Therefore, as architects, we are accustomed to building upon the work of our predecessors. Knowing this, it should encourage us to disregard the idea of having our designs as final destinations, and creating hermetic buildings that must coincide with the artist impression. Instead, we could focus on designing buildings that behave like intermediate, open-ended stations, anticipating the next change for the next generation.

To obscure – how can invisibility create value by liberating architecture from its obsession with image?

A reserved attitude with regard to modern architecture seems to be getting the upper hand when we are dealing with the act of building in an historical context. Above ground, accompanied by a motto from the 1985 movie – Back to the Future, creating new buildings that mimic the old seems to be the safest choice for architects. However, as we shift our focus from the visible to the invisible that is hidden behind a historical facade, it is there that we can find the space for unconventional forms of architecture, interior, programme, and use. In a historical city like Amsterdam, it is underground – free from any form of nostalgia and historic baggage – that there appears to be unexpectedly much room for experimenting architecture; underground stations and bicycle parking facilities are like modern-day cathedrals, winning the most prestigious awards in architecture.

To reconfigure – how can we frame our built environment as a temporary configuration of material components, waiting for a new arrangement?

New buildings are often designed for a specific purpose with a specific form. The building elements (columns, floors, walls, screws) are permanently fixed during the construction process to serve the end result, with the perception that the whole will be worth more than the sum of its parts. But what if the building, the whole, no longer serves its purpose? What then of the remaining value of its parts? Reusing old building materials has a long tradition. In the so-called spolia in Greek and Roman architecture, parts of demolished buildings were used again in new constructions. Perhaps, we ought to explore what a new type of architecture could look like – in which the building parts and their reusability will define the whole.

To repurpose – how can architecture be redefined by the act of giving new purpose to existing spaces?

There are many definitions of architecture. One might be that form and function, space and use, are so in line with each other that the way they are combined, results in a surplus value for both the space and its use. Translating this into the case of reuse, a new dialogue emerges between an existing space and its new function. If architects are experts in conceiving the best possible space for a specific use, they, too, should be just as much experts in conceiving the best use for an existing space. Redefining a building from its function, is very much redefining architecture

和使用都呈现高度的一致性。其结合方式能给空间及其用途带来剩余价值。一旦对此进行再利用,现有空间与其新功能之间就会建立新的对话。如果说建筑师是针对特定用途构思最佳空间的专家,那么他们也应该是针对现有空间构思最佳用途的专家。基于功能重新定义建筑,很大程度上是重新定义建筑本身;换言之,再定义的源头不是外部,而是内部。

密集化——如何通过增加建筑来改善现状?

几十年来,随着荷兰的日益繁荣,住宅、学校、办公场所越来越多,但这种现象只在城市,而且很少在城市内部。阿姆斯特丹的面积在"二战"后翻了一番,然而在新开发的街区,空间并不总能得到最有效的利用。城市绿化中的小型建筑原本用于人们聚会,结果,这种过剩的公共"空间"不仅没有达到预期的效果,甚至不久就变成了公共"空白"。事实上,这样的街区可以通过在现有建筑之间增加建筑,实现密集化来解决。这种做法也证明了,解决根本问题有时也有助于解决其他问题,我们可以由此创造出更丰富的多样性、更多的社区以及更好的品质。

复制——我们如何通过复制已有事物,实现品质高于原创?

建筑是艺术与工艺的结合。后两个学科中,照抄已有的设计或在设计中考虑量产,都不是什么稀奇的事情。然而,在建筑学科中,即使已经竣工,看得到规划和蓝图,复制一栋建筑仍属于"一事无成"。这就像我们想要不断设计出新的"比利",但事实上,我们应该先学会如何抄得漂亮。放下自我,尊敬前人的作品,我们作为建筑师才能重新回到自己的实践道路,并把自己的作品能被复制视为一种赞赏。

叠加——叠加新旧结构如何能够实现共利?

共生(symbiosis)这个希腊语单词意为"生活在一起",它是一种交互形式的共存。与建立在平等基础上的和谐共处不同,共生显示出,正是生物之间的差异,增加了彼此的重要性和依赖性,牛和啄牛鸟的关系便是一例。在建筑学中,则意味着在旧层的上方或下方叠加新层(可以是建筑或功能),让新旧层都能获得品质的创造和提升。

再想象——如何通过改变观点来改变现实?

没有绝对的现实。每个人对建成环境的体验都是独一无二的。这其中物理空间只起了很小的作用,更大的一部分是基于个体精神和文化的联想。这些联想,一方面是基于记忆,另一方面是基于期望。作为建筑师,我们应该认识到这一事实,尝试理解不同的人对一个建筑、一片广场、一条街道的反应。这些研究有助于我们设计出这样的建筑:它不仅允许多种解释的存在,并且能够激发出更多的解释。

(再)启动——何时才有理由忽略或抹去已经存在的事物,简单地从头开始?

在审视当下之后,人们可能会得出这样的结论:建成环境中的内容已经不再适合,需要重建。拆除和重建可能是改善现状的正确方法,但有时事实证明,过去只是一纸空文。荷兰的大部分土地都源于对水的征服,整个城市都建在排干的湖泊或内海之上,不存在可供参考的文脉。在这种情况下,就可以依循荷兰的传统,建造充分体现设计灵感的概念建筑。

itself, not from outside but from inside.

To densify – how can an existing situation be improved by adding buildings?

For decades, the increase in dwellings, schools, and offices as a result of growing prosperity in the Netherlands, happened only to the existing city, and rarely inside it. Amsterdam's surface area doubled after the Second World War and in the newly developed neighbourhoods, the available space was not always utilized in the most efficient or effective manner. Buildings were placed as objects in green zones where people could meet. As a result, this surplus of public "space" did not have the desired effect and before long, it became public "void". Densifying such neighbourhoods – by means of building among the existing buildings – is a good example of how one can sometimes solve other problems by solving an initial problem. This way, we create more diversity, more community, and hence, more quality.

To copy – how can we let quality prevail over originality by copying what is already there?

Architecture is a combination of art and craftsmanship. In both disciplines, the copying of an existing design or the thinking of mass production when developing a new design, is not something uncommon. However, in the architecture discipline, simply copying a building is still "not done", even though upon completion of a building, its plans and blueprints are there for the taking. It is rather like wanting to design a new "Billy" over and over, when in fact, we ought to embrace the art of making a good copy. In this way, by setting aside our ego and expressing esteem for the work of our predecessors, we – as architects – can rehabilitate the method of our own practice and regard a copy of our own work as a compliment.

To overlay – how can existing spatial structures overlay with new ones in a manner that each layer benefits?

Symbiosis, the Greek word for "living together" is an interactive form of coexisting. Unlike living in harmony which is based on equality, with symbiosis it is the differences between organisms that increase their importance and dependence for one another, as in the case of the ox and the oxpecker. In architecture, this means that the new layer, building, or function, above or beneath an older layer, enhances and yields more quality for both the old and new layer.

To reimagine – how can we change reality by changing our viewpoint?

There is no absolute reality. The experience with the existing built environment of each individual is unique – the physical space itself plays only a small part, and for a much larger part, it is based on the mental and cultural associations of the individual. These associations, on the one hand, are based on memories, and on the other, expectations. As architects, we should recognize this fact and try to fathom how different people react to a building, square, or street. From these studies, we can design buildings that not only allow for multiple interpretations, but also actively encourage them.

放弃——我们何时可以接受新的建筑或行动无益于改善现状或解决某个问题?

最后的行动是放弃。这是建筑师需要付出最多心力和控制力的行动,它超越了以上10个行动。在作为资本主义和消费主义中心的时代广场,超大广告牌上显示着珍妮·霍尔泽的那句"保护我免受欲望的侵蚀"。这是对我们自己的警告。在这个碳排放无处不在、有限资源被大量消耗的时代,或许这句话也可以提醒建筑师,让他们收起肤浅的偏好,不必一味追求创造新建筑、生产更多终将成为垃圾的事物。

除了上述10(+1)种方法外,当然还有其他源于不同观点、建筑文化、可持续实践的方法。设计不一定局限于其中一种;相反,最有趣的项目往往涉及多种行动。可以采用这样的混合模式:在这里打破某个障碍,在那里建造某些新事物,模仿某个空间,在原有建筑之上建造,拆除某个部件,或者索性什么都不做。

好的结局?

撰写这篇文章时,荷兰,特别是阿姆斯特丹,建筑业的情况与2012年相比已经完全逆转,经济也在回升。近来,开发商开始再次寻找房地产的商机,工地上又传出了施工作业的声音,建筑事务所依然加班加点地工作,学院里涌入了大批来自世界各地、充满热情的学生。可以说,"一切照旧"。这展现了荷兰建筑的韧性,也反映了它的机会主义。但别忘了,真正的社会问题悬而未决,建筑师的社会角色一如既往地飘摇不定。

不仅是过去不景气的七年,即使在建筑活动频繁的繁荣年代,我们建筑师也应该更多地致力于已有的事物,与那些已然置身其中的人们合作,充分利用剩余的自然资源。建筑师还需要很多:更多的参与,更多的创造,更多的智慧。我们需要这一切,以应对下一个"泥沼"的到来。

参考文献:
1. 罗萨·科尔茨森瑞吉特,罗伯特·C. 克卢斯特曼,建筑师的空间:2008-2018 荷兰建筑师地位变化之探究 [Z],阿姆斯特丹大学城市研究中心(GPIO),2018
2. 闵希豪森男爵是德国作家鲁道尔夫·埃里希·拉斯伯在其著作《吹牛大王历险记》(1785)中虚构的一个贵族人物。创作原型是贵族希罗尼姆斯·卡尔·弗里德里希·冯·闵希豪森,他曾在俄国军队服役,对抗土耳其人,他口中荒诞不经的故事就与此有关,至今民间还有口述或文本形式的记载。其中一个故事便是闵希豪森在陷入泥沼时,拉着自己的辫子,把自己和马都救了出来。

To (re)start – when is it justified to simply start from scratch and ignore, or erase what already exists?

After carefully reading what is present, one may come to the conclusion that what is written in the built environment does not fit anymore and needs to be rewritten. Demolition and building anew can also be the right method for improving an existing situation. And sometimes, as it turns out, nothing was written before. Much land in the Netherlands are conquered from the water – entire cities have been built on the bottom of drained lakes or an inland sea, without any referential context. In those cases, one can build relying on the strongly developed Dutchtradition in conceptual architecture, where everything leading to the design is to be found within the project itself.

To abstain – when do we accept that a new building or action does not contribute to the improvement of its existing situation per se or solves a particular problem?

The last action is abstaining from action. One that requires the greatest mental effort and control as an architect, and it surpasses all the above 10 actions. "Protect me from what I want" was a statement by Jenny Holzer on a large billboard in Times Square, the center of capitalism and consumerism. It is a warning to ourselves. During these times of carbon emissions and large-scale use of finite resources, the same phrase may serve as a reminder for architects to temper their frivolous tendency towards new buildings and producing more which in the end turn into waste.

In addition to the 10 (plus 1) methods mentioned above, there are certainly a diverse array of alternative methods that originate from other perspectives, building cultures and sustainable practices. Hence, a design does not always have to be limited to just one method. On the contrary, the most interesting projects often involve multiple actions. A hybrid approach provides choices like – breaking through an obstacle here, building something extra there, imitating a space, building on top of this building, demolishing a part, or doing nothing at all.

A Happy End?

At the time of this writing, the situation in the Dutch construction sector – especially in Amsterdam – is completely reversed compared to the situation in 2012. The economy has been back on the rise. For some time now, once more, developers are looking for real estate opportunities to make money, the sound of construction by the builders on site can be heard, architectural firms are working overtime, and the academy sees a great influx of enthusiastic students from all over the world. "Business as usual" one could say. It not only illustrates the resilience but also the opportunism in Dutch architecture. Yet, the real societal questions remain unsolved, making the role of the architect in society as unstable as before.

Not just in the 7 lean years, but especially in the prosperous years with high construction activity, we architects should work more with what is already there, with those who are already there, and with what we still have left – in terms of natural resources. Much more is required from an architect. We need more involvement, more creativity, more intelligence – and we will need it all for the coming of the next mire.

References:
1. Rosa Koetsenruijter and Robert C. Kloosterman, Ruimte voor de Architect: Een onderzoek naar de veranderingen in positie van architecten in Nederland 2008-2018, Centre for Urban studies/GPIO, University of Amsterdam (UvA), 2018, p. 12
2. Baron Munchausen is a fictional nobleman created by German writer Rudolf Erich Raspe in his book Baron Munchausen's Narrative of his Marvellous Travels and Campaigns (1785). The character is based on a real nobleman, Hieronymus Carl Friedrich von Münchhausen, who served in the Russian army in the fight against the Turks, about which he told very tall tales, which still continue to live on as folk tales in oral and written tradition. In one of the tales Munchausen, when caught in a swamp, rescued himself and his horse by lifting them both up by his ponytail.

Jarrik Ouburg is one of the partners of HOH Architecten, based in Amsterdam. He is a professor of Architecture at the Amsterdam Academy of Architecture, where he headed the Architecture department from 2012 until 2016 and initiated the research program, Tabula Scripta. The book *Rewriting Architecture, 10+1 Actions for an Adaptive Architecture: Tabula Scripta* has been published by Valiz Publishers in 2020.

贾里克·奥伯格是HOH建筑师事务所的合伙人之一，该事务所总部位于阿姆斯特丹。他还是阿姆斯特丹建筑学院的建筑学教授，从2012年到2016年担任建筑系主任，发起了研究项目Tabula Scripta。专著《重写建筑，适应性建筑的10+1行动：Tabula Scripta》已于2020年由Valiz出版。

Robert A.M. Stern Architects (RAMSA), Rijnboutt
Beurspassage – Oersoep
Amsterdam 2012–2016

罗伯特·A.M.斯坦恩建筑师事务所，里恩博特
展览之廊-奥索普
阿姆斯特丹 2012–2016

Credits and Data
Project title: Beurspassage – Oersoep
Client: TOP Vastgoed planontwikkeling B.V., Bouwinvest Development B.V.
Location: Damrak / Nieuwendijk, Amsterdam, the Netherlands
Design: 2012
Completion: 2016
Architect: Robert A.M. Stern Architects (RAMSA), Rijnboutt
Design team: Paul Whalen (RAMSA), Frederik Vermeesch (Rijnboutt)
Project team: FiMek estate B.V. (project management), IMd raadgevend Ingenieurs (construction), Huisman & van Muijen installatieadviseurs (installations), Peuts B.V. (fire regulations), Aboma (safety and fire regulations), Bouwbedrijf M.J. de Nijs en Zonen B.V., Dura Vermeer Groep NV (contractor), Arno Coenen, Iris Roskam, Hans van Bentem (artists)
Project area: 22,000 m²

Sauntering in contemporary nostalgia

Amsterdam's Damrak recently acquired a contemporary version of the 19th-century arcade – a public space has been given back to the sauntering crowds. Primark, a clothing chain, commissioned the renovation of the large surrounding block. They wanted the biggest possible retail surface and this gave the architect, Rijnboutt, a reason to relocate the existing arcade to the north side of the project site. Following this simple but effective spatial intervention, the arcade was taken on by Arno Coenen and Iris Roskam, both famous for their Horn of Plenty ceiling artwork in Rotterdam's Markthal (2014, *a+u*17:04).

Together with Hans van Benthem, they used mosaic and mirrors to create an artwork with images of objects that were found in the canals, from fish to rust bicycles. This "primordial soup" gives a new twist to the urban arcade which, owing to its color scheme and design, has an almost nostalgic air about it. While in Paris, 19th-century arcades defined a new type of public space where the flâneur could experiment with observing throngs of new city-dwellers; here, it is the nostalgic space that is exploited in the vocabulary of forms – from the warm and subdued colors to the traditionally crafted terrazzo floors and classic light fittings. It appeals to the yearning for a time when people had no inkling of what global capitalism would entail in terms of homogenization and commercialization of space. This arcade is redolent of a culture in which public space still gave expression to the collective dream of a modern metropolis in the making.

pp. 94–95: Interior view of the arcade. Opposite: View of the building along the streets after restoration. Photo by Rijnboutt. This page, above: Exterior view of the building before renovation. This page, below: Photo of the arcade before renovation. Photos on pp. 94–101 by Kees Hummel unless otherwise specified.

第 94-95 页：拱廊内部。对页：整修后的建筑沿街外观。本页，上：大楼改造前外观；本页，下：改造前的拱廊。

Passageway ceiling artwork (scale: 1/800)／走廊天花板的艺术作品（比例：1/800）

Passageway east wall artwork／走廊东侧墙壁的艺术作品

Passageway west wall artwork (scale: 1/800)／走廊西侧墙壁的艺术作品（比例：1/800）

Passageway floor artwork (scale: 1/800)／走廊地面的艺术作品（比例：1/800）

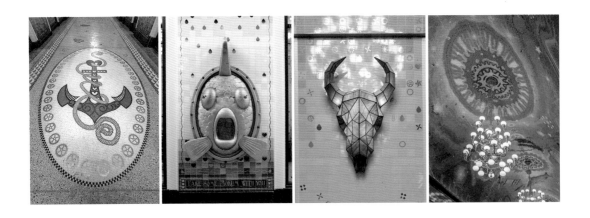

98

在当代怀旧中漫步

阿姆斯特丹的达姆拉克大街最近开放了一个当代版的19世纪拱廊,作为公共空间被重新归还给漫步的人群。普利马克服装连锁店委托设计师对周围的大型街区进行改造。他们想要获得最大的零售店面,这给了建筑师里恩博特一个将现存拱廊迁至项目场地北侧的理由。遵循简单而有效的空间干预原则,该拱廊由阿诺·科恩和艾里斯·罗斯卡姆接手,二人均以鹿特丹市集广场"丰裕之角"天花板的艺术创作而闻名(2014,《a+u 17:04》)。

他们与汉斯·范·本瑟姆一起,用马赛克和镜子创作出这件艺术品,从鱼到生锈的自行车,包罗其中的形象皆源自在运河中打捞的物品。这里的一件作品"原生浆液"(地球上生命出现之前存在的一种液态物质)给城市拱廊带来了新的扭动感,其配色和设计充满怀旧气息。在巴黎,19世纪的拱廊定义了一种新型公共空间,信步闲游者可以在这里观察三五成群的新城市居民。所有描述形态的词汇都可以汇集在这个怀旧空间——从温暖柔和的色彩到传统的水磨石地板和经典的灯具。它引起人们对这样一个时代的向往,一个摆脱全球资本主义带来的空间同质化和商业化影响的时代。这个拱廊充溢着这样一种文化之馥:至今,公共空间依然表达着人们对未来现代大都市的共同梦想。

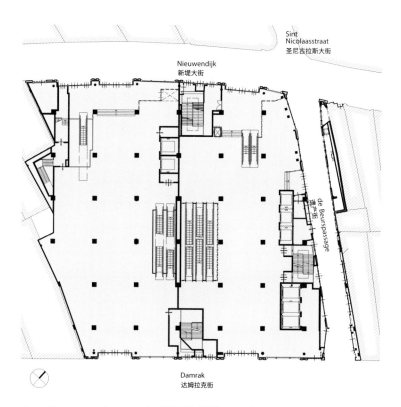

Ground floor plan (scale: 1/700)／一层平面图(比例:1/700)

Opposite, below: Artwork, Oersoep. From left to right: Mosaic art on the floor. Photo courtesy of Frank Hanswijk. Big ornament in the form of a fish. Glass sculpture in the form of an ox. Glass mosaic on the ceiling.

对页,下:奥索普艺术作品。从左至右依次为:地面马赛克艺术,大型鱼形装饰,牛形玻璃雕塑,天花板上的玻璃马赛克。

Elevations of arcade (scale: 1/500) ／拱廊立面图（比例：1/500）

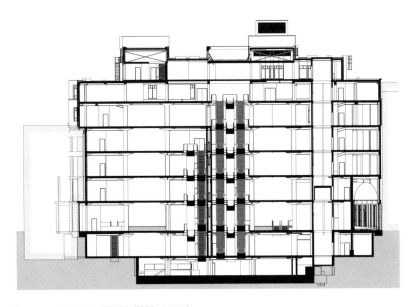

Section (scale: 1/700) ／剖面图（比例：1/700）

Opposite: Exterior view from the east. The facade is refurbished to blend with the surrounding cityscape. The arcade work was part of the full renovation of the building.

对页：东面外观。立面经过翻新，与周围的城市景观融为一体。拱廊工程是整栋大楼改造的一部分。

Superuse Studios
BlueCity Rotterdam
Rotterdam 2015–2017

超级利用工作室
鹿特丹蓝色城市
鹿特丹 2015–2017

For over 20 years, Rotterdam-based Superuse Studios has operated out of a conviction that the building sector should deal far more responsibly with available materials. Its belief in the need to utilize waste that streams from other branches of the industry means that the firm, by definition, works outside the prevailing parameters in the construction industry. For this approach to succeed, it is necessary to link up with like-minded businesses, preferably physically on a single site. Such an opportunity arose with the auction of an obsolete Tropicana swimming paradise in Rotterdam. Together with ifund and Coup, they won the bidding with their plan, BlueCity 010. The redevelopment of the leisure complex to a national platform for circular entrepreneurs will take place in steps, and this started in 2016 with an office wing. Superuse Studios used reclaimed timber frames for these offices. Their size meant that they could only be positioned at an angle, with the result that the frames offer views into the offices while screening them from the corridor. Currently, the redevelopment of the basement into workspaces, laboratories, ateliers, and offices is taking place.

Blue City is much more than the redevelopment of an old building. It is a form of area development in which the tenants become part of a recycling system that will evolve over the years. The large hall, once home to the famous wild water ride, will become a market hall. This will be an explicit link with Rotterdam, a place where experiences and knowledge about sustainability and the circular economy can be exchanged between businesses, designers, civil servants, and citizens.

Superuse Studios

A swimming pool full of possibilities, that is BlueCity. A playground for circular companies, where we race with brains, balls, guts, and fun into an economy where waste does not exist. In BlueCity, you get a portion of radical disruption with your coffee and together, we could realize a world in which waste is valuable. It is a place where innovative companies link their residual flows together. What waste is for one, is input for the other. In this way, together, we create an exemplary city for the circular economy.

Blue City is funded with an enormous amount of guts. It is a city with a new sustainable business model in which waste is seen as raw material. A city where beeswax goes to the furniture maker and new products are made from plastic waste. If this works, we will never have waste again. We will be happy with coffee grounds and mushrooms, and it will grow in size with time, achieving even more impact. BlueCity is not about demolishing, but about harvesting.

COUP, real estate development partner of BlueCity

---

p. 102, above: Photo taken before renovation. The office floor was once a disco hall. Photo courtesy of the architect. p. 102, below: Photo of the swimming pool taken before renovation. Photo courtesy of the architect. p. 103: Office space in BlueCity Rotterdam. The swimming pool and its restaurant has been renovated into a platform for entrepreneurs. Photo by Frank Hanswijk. Opposite: Aerial view from the south. Photos of p.105, p.106 by Denis Guzzo.

第 102 页，上：改造前的景象。如今的办公空间曾经是一个迪斯科舞厅；第 102 页，下：改造前的游泳池。第 103 页：鹿特丹蓝色城市内的办公空间。游泳池及餐厅被改造成一个企业家平台。对页：南面鸟瞰图。

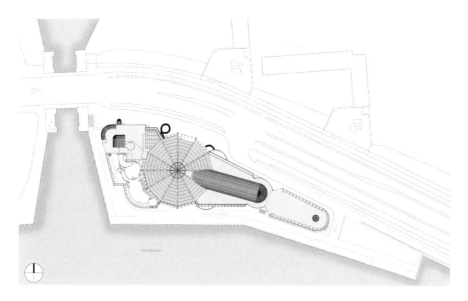

*Site plan (scale: 1/2,500／*总平面图（比例：1/2,500）

Credits and Data
Project title: BlueCity Rotterdam
Client: BlueCity
Location: Rotterdam, the Netherlands
Design: 2015
Completion: 2017
Architect: Superuse Studios
Design team: Floris Schiferli, Maartje Kool, Jeroen Bergsma, Frank Feder
Project team: Climatic Design Consult (climate technology), Engie (installation technology), IOB (construction and safety consultant), ROOT (BIM experts), Oogstkaart B.V. (reused materials), BIK Bouw, Mostert B.V. (construction)
Project area: 1,440 m² (gross floor area)

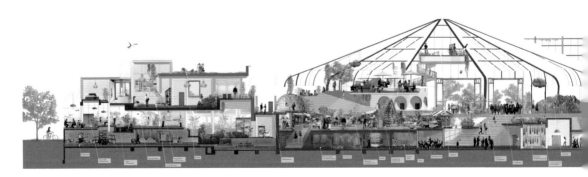

Perspective section (scale: 1/800) ／剖面透视图（比例：1/800）

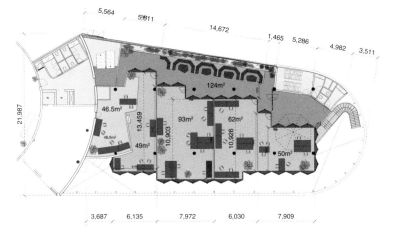

First floor plan of the office space／办公空间二层平面图

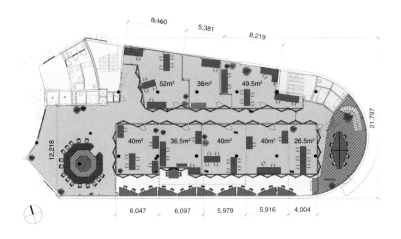

Ground floor plan of the office space (scale: 1/500)／办公空间一层平面图（比例：1/500）

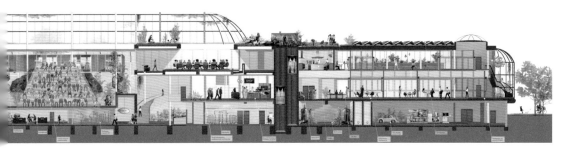

总部位于鹿特丹的超级利用工作室20多年来一直坚信建筑行业应对现有材料的处理承担起更多的责任。他们坚信对行业其他分支产生的废物进行再利用是必要的，这也意味着该工作室的工作显然超出了建筑行业的普遍标准。为了使这种方法取得成功，他们认为有必要与志同道合的企业建立联系，最好是能进行合作。这样的机会终于在鹿特丹老旧的"热带花园游泳天堂"被拍卖时出现了。他们与ifund和Coup联合，以蓝色城市010方案中标了该项目。这个把休闲综合体重建为全国性循环利用的企业家平台的项目由2016年启动的侧翼办公楼拉开帷幕，并将逐步展开。超级利用工作室将回收的木材框架用于这些办公。它们的尺寸意味着它们只能以一定的角度被组合放置，这些框架既能为办公室引入景观，又能在走廊一侧为其提供遮挡。目前，地下室正在被重新开发为工作空间、实验室、艺术工作室和办公室。

蓝色城市远不止是对一座老建筑的改建，它更是一种区域发展形式，使租户成为回收系统的组成部分，该系统也将随着时间的推移逐步发展。曾经装载着著名"疯狂水上滑道"的大型娱乐中心将变成一个市集大厅。这也将成为该项目与鹿特丹最直接的联系，因为在鹿特丹，企业、设计师、公务员和公民之间都可以交流有关可持续性和循环经济的经验和知识。

超级利用工作室

蓝色城市是一个充满可能性的游泳池。这里是循环利用企业的乐园，在这里我们用智慧、胆量、勇气和趣味向着一个不存在废物的经济体全速行进。在蓝色城市，您的办公室内会有一部分地面彻底地与周围分裂，那片地面是用您的咖啡渣填充而成的，由此我们可以共同实现一个废物即资源的世界。在这里，创新型公司将其剩余流量连接在一起。于一些人是废物，于他人则是资源。通过这种方式，我们将共同打造一个循环经济示范城市。

蓝色城市是以极大的勇气建立起来的。这是一个具有新型可持续商业模式的城市，这里的废物被视为原材料。在这里，蜂蜡为家具制造商所用，新产品产自废弃塑料。如果这行得通，那我们将再无废料。我们对咖啡渣地面和蘑菇种植感到满意，并且随着时间的推移，它们的规模不断扩大，从而产生更大的影响。蓝色城市，并非破坏，而是一种收获。

COUP，蓝色城市地产开发合作伙伴

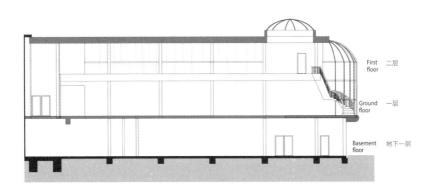

Section of the office space (scale: 1/500)／办公场所剖面图（比例：1/500）

p. 106: Photo of the pool with its water drained. Presently, it is to be renovated into a market hall. Opposite, above: Office space on the first floor. The round sofas along the corridor are recycled from the old swimming pool. Photo by Denis Guzzo. Opposite, below: Office space on the ground floor. Recycled timber frames are used for the windows giving the office a unique zigzag spatial experience. Photo by Frank Hanswijk.

第106页：排空水的游泳池。如今，它将被改造成一个市集大厅。对页，上：位于二层的办公空间。沿走廊排布的圆形沙发是对原游泳池沙发的再利用；对页，下：位于一层的办公空间。回收的木框架被用作窗框，为办公室带来独特的"Z"字形空间体验。

Neutelings Riedijk Architects
Deventer City Hall
Deventer 2010–2016

诺特林斯·里迪克建筑师事务所
代芬特尔市政厅
代芬特尔 2010–2016

The new city quarter unites the old city hall with a new city office at "Grote Kerkhof" and "Burseplein" in Deventer. The City Hall will be a public building that houses all of Deventer's employees. The city council and all municipal services for citizens and companies will be stationed in the new building. Frequented by visitors and users of the City Hall, the formerly abandoned part of the city once again becomes lively. It is one of the most sustainable government buildings in the Netherlands.

A Maze of Yards and Squares
The City Hall's design seamlessly blends in with Deventer's typical urban tradition of interconnected gardens, squares, and doors. The building is organized around 2 gardens: one that encloses the old mayor's house, and the other, a covered inner yard where citizens, visitors, and employees can meet. Next to the buildings in Polstraat and Assenstraat, there are 2 green city gardens that join these adjacent buildings.

Accessibility and Circulation
The City Hall creates a link between the city's 2 most important squares, the "Brink" and the "Grote Kerkhof". Perpendicular to the City Hall is a walking route that connects the inner city to the Ijssel river through streets and alleyways. All buildings that make up the new City Hall are connected to each other on the ground level. The main route circles around the Mayor's Yard and joins the historical city hall to the new city office.

Fit in the Environment
The front building at "Grote Kerkhof" consists of 2 layers with a sloped roof and strong vertical lines, much like its neighbor "de Hereeniging" and the city hall on the left. This continuity makes the front building fit in with its surroundings. By having a receding roof, the facade appears to separate itself from the roof-plane. This creates a city balcony and turns the facade into a screen at "Grote Kerkhof". The main building has 3 layers, like most houses surrounding it. The long vertical windows blend in with the inner city's architecture and allow light to cast deep into the building.

The City Hall has a roof surface consisting of alternating mansards and green flat roofs, thus creating a varied skyline that blends in with its environment. The mansards provide space for stylish inner courts in the attics. On the side of the Mayor's Yard, is a roof garden that offers a view of Deventer's inner city.

Facades
Brick volumes and filigree wooden frames alternate on the facades. The stone facades consist of large size brickwork, while the filigree facades consist of metal grids inserted in oaken wood frames. The decorative grid covers up edges of the floor. The grids and fences are designed in collaboration with the artist, Loes ten Anscher.

Sustainability
The City Hall optimizes the use of natural resources as much as possible. The use of sunlight, rainwater, and water from the Ijssel river increases the building's sustainability by 25%, as compared to similar offices in the Netherlands. The many windows, with their natural light and air circulation generated by high ceilings, make the office an agreeable working space.

pp. 110–111: View through the courtyard into the west entrance of the city hall. The new city hall was constructed to complete the spaces and connectivity between existing buildings. Opposite, above: West exterior view. Opposite, below: Northeast exterior view. Photos on pp. 110–121 by Scagliola Brakkee, courtesy of Neutelings Riedijk Architects.

第110-111 页：穿过庭院看到的市政厅西入口。新市政厅的建造是为了完善现有建筑之间的空间和连接。对页，上：西面外观；对页，下：东北面外观。

代芬特尔新市政厅将老市政厅与位于"格罗特·克尔霍夫"和"伯塞普林"的城市办公空间结合在一起。新市政厅将是一座能容纳代芬特尔所有职工的公共建筑。市议会以及为市民和企业提供的所有市政服务都将设在这座新建筑中。由于市政厅工作人员和访客时常光顾,这片曾被人遗忘的市中心区重新恢复了活力。它也是荷兰最具可持续性的政府建筑之一。

迷宫般的庭院和广场

将花园、广场和门互相连通,市政厅的设计完美地融入了代芬特尔典型的城市传统中。该建筑围绕两个花园组织空间:其中一个花园环绕着前市长公馆,另一个是带顶棚的室内庭院,市民、访客和职员可以在这里会面。在波尔斯特拉特和阿森斯特拉特的建筑物旁,两片绿色城市花园将这些彼此相邻的建筑连接在一起。

可达性和交通流通

市政厅在城市最重要的两个广场"布林克"和"格罗特·克尔霍夫"之间建立了联系。垂直于市政厅的是一条步行路线,它通过街道和小巷连接起了内城与艾瑟尔河。组成新市政厅的所有建筑都在地面层相互连通。主要路线环绕市长公馆,并连通了历史悠久的市政厅和新市政办公空间。

融入环境

"格罗特·克尔霍夫"广场前楼是一座带坡面屋顶的二层建筑,其立面像毗邻的老市政厅和"de Hereeniging"社交俱乐部一样,有着强烈的纵向线条感。这种连续性使前楼与周围环境融为一体。其屋顶向后退让,使立面看上去似与屋顶分离。而由此形成的城市露台将建筑立面转变为"格罗特·克尔霍夫"的一道屏障。像周围的大多数房屋一样,主体建筑为三层。纵向的高窗使其毫不违和地融入城内建筑,并使光线得以深入建筑内部。

市政厅的屋顶由交替的孟莎式屋顶(又称复折式屋顶,是两折的双坡顶)和绿色平屋顶组成,由此形成了一个与环境一致且富有变化的天际线。其复折式屋顶设计为屋顶之下的现代风格的内院提供了空间。在市长公馆的一侧,是一个可以一览代芬特尔内城景色的屋顶花园。

立面

大量砖砌和浮雕装饰的木质框架在立面上交替出现。石造立面由大尺寸的砌块组成,而浮雕装饰的立面则由镶嵌着金属网格的橡木框架组成。装饰性的网格沿地面边缘覆盖着建筑。这些网格和围栏是与艺术家卢斯·特恩·安舍尔合作设计的。

可持续性

新市政厅最大限度地优化了对自然资源的利用。相较于荷兰类似的办公空间,该建筑对阳光、雨水以及艾瑟尔河水的利用使建筑的可持续性提高了25%。不计其数的窗户带来的自然采光,加上由高敞的天花板产生的空气流通,为办公室创造了舒适的工作空间体验。

pp. 114-115: View of the city hall's atrium. Opposite, above: Restaurant on the first floor. Opposite, below: Reception on the ground floor.
第 114-115 页:市政厅的中庭。对页,上:二层的餐厅;对页,下:一层的接待处。

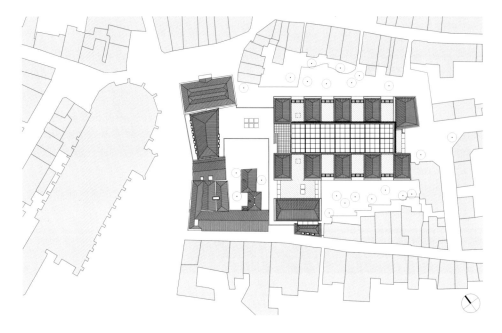

Site plan (scale: 1/2,500)／总平面图（比例：1/2,500）

Credits and Data
Project title: Deventer City Hall
Client: City of Deventer
Location: Grote Kerkhof 1, Deventer, the Netherlands
Design: October 2010
Completion: April 2016
Architect: Neutelings Riedijk Architects
Project team: ABT, Adviesbureau voor Bouwtechniek B.V. (architectural engineer), Aronsohn Constructies Raadgevende Ingenieurs B.V. (structural engineer), DGMR Raadgevende Ingenieurs B.V. (building physics), Hiensch Engineering B.V. (building services), Basalt Bouwadvies B.V. (cost consultant), Neutelings Riedijk Architecten, Atelier PRO (interior design), OTH Architects (council meeting room interior design), Bosch Slabbers tuin en landschapsarchitecten (landscape architect), BAM Utiliteitsbouw Regio Noordoost (general contractor), Design 'Deventer Framework': Loes ten Anscher (art integration)
Project area: 19,500 m² (new building), 4,500 m² (restoration)

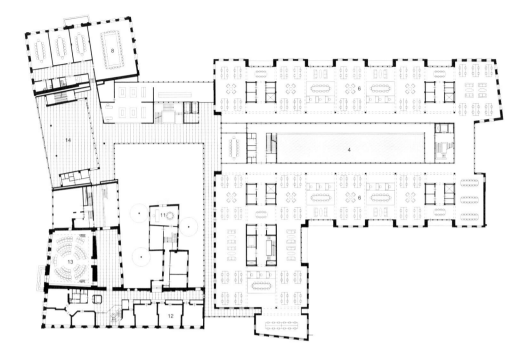

First floor plan／二层平面图

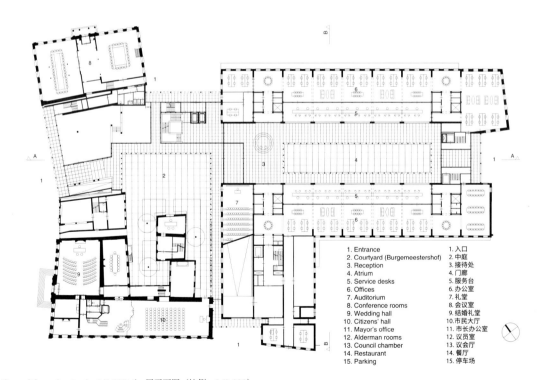

1. Entrance
2. Courtyard (Burgemeestershof)
3. Reception
4. Atrium
5. Service desks
6. Offices
7. Auditorium
8. Conference rooms
9. Wedding hall
10. Citizens' hall
11. Mayor's office
12. Alderman rooms
13. Council chamber
14. Restaurant
15. Parking

1. 入口
2. 中庭
3. 接待处
4. 门廊
5. 服务台
6. 办公室
7. 礼堂
8. 会议室
9. 结婚礼堂
10. 市民大厅
11. 市长办公室
12. 议员室
13. 议会厅
14. 餐厅
15. 停车场

Ground floor plan (scale: 1/1,000)／一层平面图（比例：1/1,000）

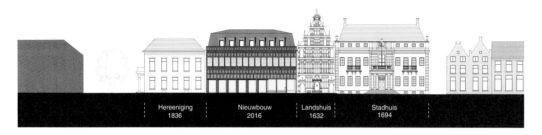

North elevation (scale: 1/1,000)／北立面图（比例：1/1,000）

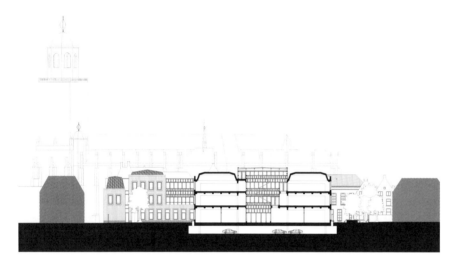

Section B／B剖面图

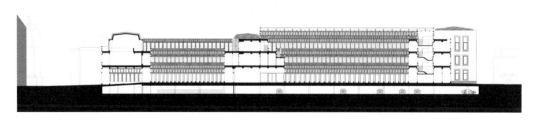

Section A (scale: 1/1,200)／A剖面图（比例：1/1,200）

This page: Facade details. Fingerprints of its citizens are used as motifs and applied repeatedly on the inside and outside the building. Photos on pp. 110–121 by Scagliola Brakkee, courtesy of Neutelings Riedijk Architects.

本页：立面细部。市民的指纹被用作装饰图案，并在建筑内外被反复使用。

CIVIC Architects
LocHal Library
Tilburg 2015–2018

城市建筑事务所
劳克哈尔图书馆
蒂尔堡 2015–2018

pp. 122-123: View from the main staircase on the first floor. The 90×60m wide and 15m high locomotive hangar was rehabilitated into a local public library. This page, above: Aerial view of the site when it was still being used as a locomotive hangar. This page, below: Interior view of the locomotive hangar before construction. Phots on this page courtesy of the architect. Opposite: View from along the corridor on the first floor. A new frame is inserted into the existing frame to provide circulation within the buildling. Photos on pp. 122-133 by Stijn Bollaert unless otherwise specified.

第 122-123 页：于二层主楼梯上看到的景色。这个长 90 米、宽 60 米、高 15 米的火车停放库已被改建为当地的公共图书馆。本页，上：该场地仍被用作火车停放库时的鸟瞰图；本页，下：改建前的火车停放库内部。对页：沿二层走廊眺望。新的结构被置入现有框架中，为建筑内部提供新流线。

The LocHal is a new public library for the City of Tilburg. The former locomotive hangar was transformed into a public meeting place that turns "the wrong side of the tracks" into a vibrant mixed-use district.

The LocHal redefines library typology in the digital age. While retaining traditional library facilities, the new library also provides ample opportunity for interaction and the creation of new knowledge. The building acts as a large covered public space, housing amenities shared by the library, arts organizations and co-working facilities. In addition to areas for lectures and public events, the building has a number of "labs" where visitors can learn new skills.

The architecture is a reinterpretation of the original late-industrial building dating back to 1932. With a height of 15 m, it is both imposing and inviting. This spaciousness is strengthened by a landscape of stairs and diagonal sightlines across the interior. Enabled by smart engineering, the library capitalizes on the existing structure, thereby greatly minimizing the amount of new structural elements. At the same time, an ingenious system of climatic zones preserves the openness of the building.

The building shapes the flow of people. The landscape of stairs leads visitors up into the building towards a gallery which allows visitors to browse books or retreat into one of the quieter reading areas. Higher up, a large balcony offers panoramic views over the city. 6 giant movable textile screens create flexible zones within the folded landscape, and improves acoustics. During the day, abundant daylight creates intricate shadow patterns. After dark, the building is turned "inside out", with its interior becoming an inviting beacon in the city.

• Locomotive hall

The hull of Nedtrain's historic Locomotive Hall forms the basis for the architecture. All traces of use have been retained. They set the key-note for the new architecture.
- Braaksma & Roos Architectenbureau

• 火车大厅

内德特兰历史悠久的火车大厅的外壳构成了建筑的基础。所有的使用痕迹都保留了下来。这些为新建筑起了关键作用。

-布拉克斯马和鲁斯建筑事务所

• Interwoven architecture

The solid architecture coincides with the hall and makes the LocHal an impressive public place: An open knowledge workshop with various labs and space for events.
- CIVIC Architects
- Braaksma & Roos Architectenbureau
- Inside Outside / Petra Blaisse

• 交织建筑

坚固的建筑与大厅相得益彰，使劳克哈尔成为一个充满活力的公共场所：一个有各种实验室和活动空间的开放式知识工作间。

-城市建筑事务所
-布拉克斯马和鲁斯建筑事务所
-内外/皮特拉·布莱斯

• Movable textiles

6 massive movable screens make it possible to divide the hall and stairwell into different zones for lectures, events and exhibitions.
- Inside Outside / Petra Blaisse
- TextielLab

• 可移动的纺织品

6块巨大的活动屏风可以将大厅和楼梯间分成不同的区域，用于演讲、活动和展览。

-内外/皮特拉·布莱斯
-纺织实验室

• Colorful life

Many different activities and targetgroups are located side by side in the hall, with different design themes: An interior that is full of diversity.
- Mecanoo
- Academy of architecture Tilburg

• 丰富多彩的生活

大厅中并排存在着不同的活动和不同的目标人群，他们都有着不同的设计主题：一个充满多样性的室内空间。

-麦卡诺
-蒂尔堡建筑学院

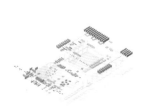

Concept diagram／概念分析图

Opposite, above: Interior view from the south entrance along the west side of the building on the ground floor. From the main staircase, visitors will find a learning laboratory on the first floor, and a glass hall on the second floor. Opposite, below: Interior view from the south entrance along the east side of the building on the ground floor. Privacy of the spaces increases as the visitors ascend past each floor.

对页，上：从一层西侧的南入口望向室内。在主楼梯上，来访者会发现位于二层的学习实验室和三层的玻璃大厅；对页，下：从一层东侧的南入口望向室内。空间的私密性随着来访者进入更高楼层而增强。

1. Entrance	14. Opinion and debate lab	1. 入口	14. 交流辩论室
2. City cafe (stand cafe)	15. Knowledge making lab	2. 城市咖啡厅（站立式咖啡厅）	15. 知识创造实验室
3. Reading places and open podium	16. Workplace seats2meet	3. 阅读区和开放式讲台	16. 工作场
4. Exhibition space	17. Concert hall	4. 展览空间	17. 音乐厅
5. Youth library	18. Kitchen seats2meet	5. 青年图书室	18. 厨房
6. Digilab	19. Conference rooms seats2meet	6. 数字实验室	19. 会议室
7. Library	20. Informal work and debate stand	7. 图书室	20. 非正式的工作和辩论台
8. Cooking lab	21. Workplaces around the workplace	8. 烹饪实验室	21. 环绕式工作空间
9. Living library	22. Science collection	9. 生活图书室	22. 科学图书收藏室
10. Meeting rooms	23. Project rooms	10. 会议室	23. 项目室
11. Office	24. Concentration workplace	11. 办公室	24. 免扰办公空间
12. Mezzanine	25. Restaurant city balcony	12. 夹层	25. 城市露台餐厅
13. Heritage lab	26. Textile walls	13. 遗产研究室	26. 织物外墙

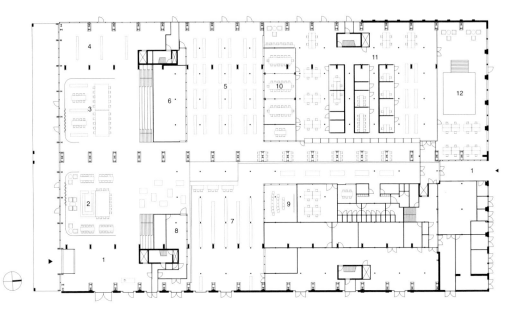

Ground floor plan (scale: 1/800)／一层平面图（比例：1/800）

劳克哈尔是蒂尔堡市一座新建的公共图书馆。曾经的火车停放库被改造成一个公共集会场所，将"铁轨的另一边"变成了一个充满活力的多功能区域。

劳克哈尔重新定义了数字时代的图书馆类型。新图书馆在保留传统图书馆设施的同时，也为互动和知识创新提供了充分的机会。该建筑作为一个带顶棚的大型公共空间，将为图书馆、艺术组织和联合办公提供便利的共享设施。除了讲座和公众活动区外，建筑内还有许多"实验室"，来访者们可以在这里学习新技能。

该建筑是对原有后工业时期建筑的重新诠释，原建筑可追溯至1932年。这座15米高的建筑，既雄伟又迷人。室内大面积的楼梯以及通透的对角线视野增强了其空间感。通过智能工程技术，该图书馆能够利用现有结构，最大限度地减少新结构的使用度。同时，巧妙的气候分区系统使建筑的开放性得以保留。

该建筑影响着人群流动。楼梯景观引导来访者上行进入建筑，并通往一个陈列区，来访者可以在这里浏览书籍或是前往更安静的阅读区。较高处的大露台使来访者可以尽情欣赏城市全景。6个巨大的可移动织物幕帘在阶梯状的景观中建立起灵活分区，并改善了传声效果。在白天，充足的日光形成错综的阴影图案；天黑后，这座建筑"由内向外"翻转，成为城市中一座引人瞩目的灯塔。

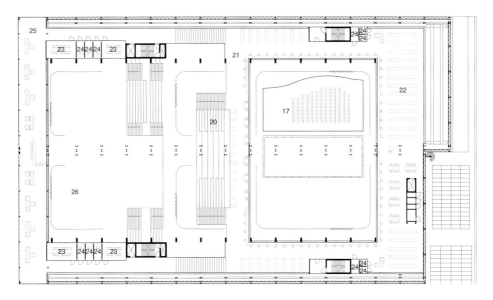

Second floor plan／三层平面图

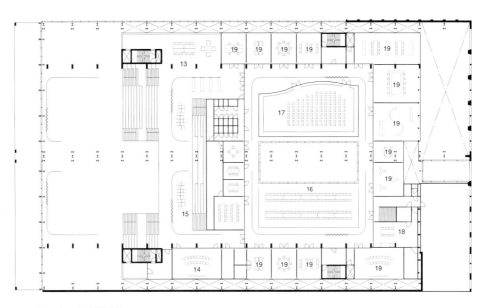

First floor plan／二层平面图

129

Credits and Data
Project title: LocHal Library
Client: City of Tilburg
Location: Tilburg, the Netherlands
Design: 2015
Completion: 2018
Architect: CIVIC Architects (lead architect), Braaksma & Roos Architectural Office (transformation and restoration)
Project team: Inside Outside / Petra Blaisse (interior and landscape design, textiles), Donkergroen (landscaping), Mecanoo (interior design library and offices), Zaanen Spanjers Architecten (octatube glass hall), Stevens van Dijck Bouwmanagers & Adviseurs (construction management), Arup (technical consultancy), Binx Smartility (lead contractor), VDNDP Structural Engineers (site architect), F. Wiggers Ingenieursbureau (structural engineering), ABT Wassenaar (building physics, acoustics, fire safety), SOM (commercial management), Linneman (BREEAM certification), Inside Outside/Petra Blaisse (textile screen – design drawings and installation), Tilburg Textile Museum (woven sections textile lab), Seilemaaker (canvas sections and installation), Gerriets France (assistant canvas sections), Theatex BV (textile screen – monofilament sections), Gerriets Gmbh (textile screen – rails and motorized system), Levtec (textile screen – installation and software)
Project area: 11,200 m²

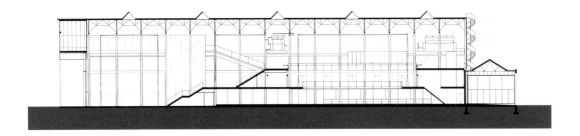

Long section／横向剖面图

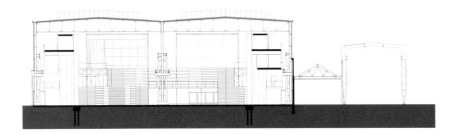

Short section (scale: 1/800)／纵向剖面图（比例：1/800）

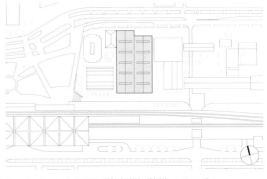

Site plan (scale: 1/6,000)／总平面图（比例：1/6,000）

pp. 130–131: North exterior view. Opposite: View from the Tilburg station platform. This page: South exterior view.

第 130-131 页：北面外观。对页：于蒂尔堡车站月台看到的景色。本页：南面外观。

Neutelings Riedijk Architects
Naturalis Biodiversity Center
Leiden 2019

诺特林斯·里迪克建筑师事务所
荷兰国家生物多样性中心
莱顿 2019

The institute's new design forms a sustainable ensemble of existing buildings and new-build with each activity housed in a specific form. The central atrium connects various parts of the institute: the existing offices and depots with the newly built museum and laboratories. The design of the atrium consists of a 3-dimensional concrete structure in the form of interlocking molecules – a lace of ovals, triangles, and hexagons. Filtered light entering through the circular windows reinforce the monumentality of the space, like a "glass crown" where scientists, staff, students and families meet.

Public functions such as the restaurant, the shop, and the exhibition hall can be found on the ground floor where passers-by can catch sight of the examinations of the last whales washed ashore. The main staircase leading up to the exhibitions resembles a mountain path – becoming narrower at the top with enough space to welcome Trix, the 66 million years old T-Rex which has been given pride of a place in the Dino Era gallery.

The exterior of the exhibition halls covered with stone blocks in horizontal layers mimics a geological structure. The variety of travertine stone used had developed natural crystals over the span of eons, creating a sparkle. The layers of stones are interrupted by friezes of white which are concrete elements designed by Dutch fashion designer, Iris van Herpen. Invited by Neutelings Riedijk Architects, she designed a total of 263 panels inspired by the natural shapes of the collection which seem to be as smooth as silk, thanks to a special technique developed for Naturalis.

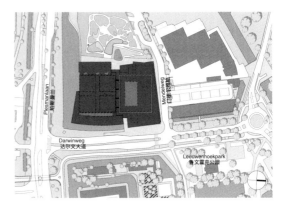

Site plan (scale: 1/5,000)／总平面图（比例：1/5,000）

Credits and Data
Project title: Naturalis Biodiversity Center
Client: Naturalis Biodiversity Center
Location: Darwinweg 2, Leiden, the Netherlands
Design: March 2013
Completion: May 2019
Architect: Neutelings Riedijk Architects
Design team: Michiel Riedijk, Willem Jan Neutelings, Frank Beelen, Kenny Tang, Guillem Colomer Fontanet, Jolien Van Bever, Inés Escauriaza Otazua, Marie Brabcová, Cynthia Deckers
Project team: ABT BV Ingenieursbureau (architectural engineer), Aronsohn Raadgevende Ingenieurs (structural engineer), Huisman en van Muijen (installation design), DGMR Raadgevende Ingenieurs (building physics), J.P. van Eesteren (general contractor), IC ULC-Kuijpers (installations contractor), Neutelings Riedijk Architects (general public area interior design), Hollandse Nieuwe (offices interior design), Iris van Herpen (artwork, concrete relief), Studio Tord Boontje (artwork, graphic), Studio Hartzema (urbanist), IGG, Bointon de Groot (cost consultant), H+N+S (landscape architect)
Project area: 20,000 m² (new building), 18,000 m² (renovation of existing building)

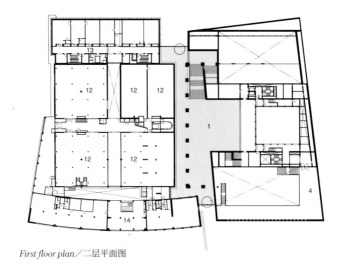

First floor plan／二层平面图

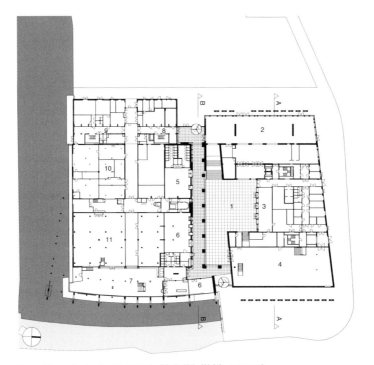

Ground floor plan (scale: 1/1,500)／一层平面图（比例：1/1,500）

1. Atrium
2. Restaurant
3. Shop
4. Exhibition
5. Cloakroom
6. Education room
7. Entrance offices
8. Entrance laboratories
9. Laboratories
10. Workshops
11. Book depot
12. Collection depots
13. Laboratories
14. Offices
15. Lobby
16. Reception hall
17. Roof terrace
18. Technical equipment room

1. 中庭
2. 餐厅
3. 商店
4. 展览厅
5. 衣帽间
6. 教室
7. 入口办公室
8. 入口实验室
9. 实验室
10. 工作坊
11. 图书储藏室
12. 收藏仓库
13. 实验室
14. 办公室
15. 门厅
16. 接待厅
17. 屋顶天台
18. 技术设备间

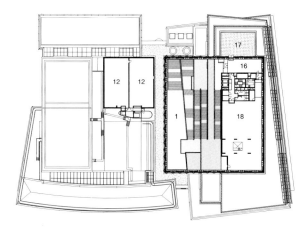

Ninth floor plan／十层平面图

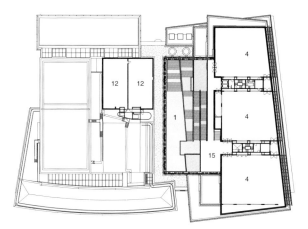

Seventh floor plan／八层平面图

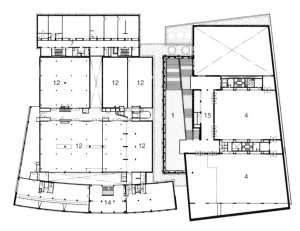

Third floor plan／四层平面图

139

p. 135: East exterior view. Photos on pp. 134–143 by Scagliola Brakkee, courtesy of Neutelings Riedijk Architects. pp. 136–137: Atrium. Visitors access the exhibition rooms on each floor through the main atrium stairs. The atrium's ceiling height is approximately 40 m high. This page, left: Interior view from the atrium looking towards the south facade. This page, right: Details on the wall are composed of laminating travertine and concrete friezes. Opposite: View of the south columns and wall cladded in wooden panels from the atrium ground floor.

第 135 页：东面外观。第 136-137 页：中庭。参观者可通过主中庭楼梯进入每层的展览室。中庭天花板的高度约为 40 米。本页，左：从中庭望向南面内立面；本页，右：墙上的细节由石灰华压板和混凝土饰条组成。对页：在中庭底层望向南面立柱和由木板覆盖的内墙。

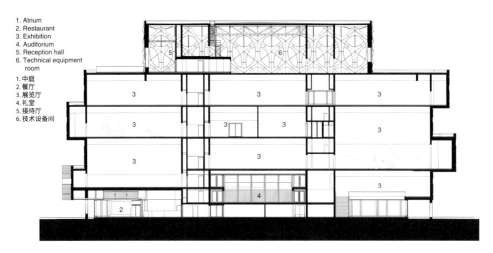

1. Atrium
2. Restaurant
3. Exhibition
4. Auditorium
5. Reception hall
6. Technical equipment room

1. 中庭
2. 餐厅
3. 展览厅
4. 礼堂
5. 接待厅
6. 技术设备间

Section A／A剖面图

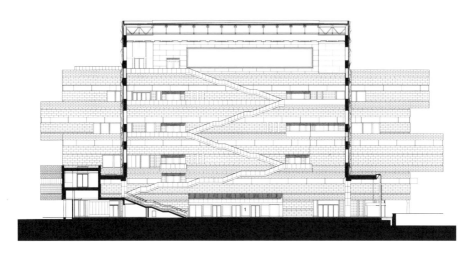

Section B (scale: 1/800)／B剖面图（比例：1/800）

Opposite: Exterior view from the west. The new building can be seen on the left, while the existing is on the right.

对页：西面外景。新建筑在左边，原有建筑在右边。

研究中心的新设计使得现有建筑与新增部分组成了一个可持续的整体，每种功能空间都有其对应的不同形式。中庭将建筑的各个部分连通，包括现有的办公室和仓库，与新建的博物馆和实验室。中庭的设计为三维的混凝土结构，就像由椭圆形、三角形和六边形组成的互扣的分子结构形态"花边"。光线透过圆窗洒向这个科学家、工作人员、学生与家庭相聚的空间，如同一顶"玻璃皇冠"，强化了空间的纪念性。

餐厅、商店和展览厅等公共功能空间均位于底层，游客可以在这里看到关于最后被冲上岸的鲸鱼的调查。

通往展览的主楼梯像是一条山路，在顶部变得狭窄，为顶层提供了足够的空间展示6,600万年前的霸王龙化石"特里克斯"。这是恐龙时代展厅中令人骄傲的展品。

展厅的外部饰材采用水平堆叠的石块，模仿了天然的地质结构。所使用的洞石在漫长的岁月中形成了天然的晶体，闪烁着耀眼的光彩。石层间的白色混凝土带状元素由荷兰时装设计师艾瑞斯·范·赫本所设计，受诺特林斯·里迪克建筑师事务所的邀请，她共设计了263块灵感来源于藏品自然形态的面板。得益于为该项目所研发的特殊技术，这些面板看起来像丝绸一样光滑。

KAAN Architects
Education Center Erasmus MC
Rotterdam 2006–2013

KAAN建筑师事务所
伊拉姆斯大学医学教育研究中心
鹿特丹 2006–2013

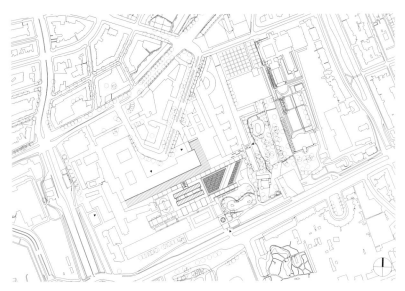

Site plan (scale: 1/10,000)／总平面图（比例：1/10,000）

The Medical Studies Center, in the academic section of Erasmus MC, is a beneficent sea of space and light under a high ceiling with lots of glass. This used to be the second level; a paved exterior courtyard originally designed to guide the user on an attractive route through the complex. The connecting space never really worked. It has now been converted into a well-used atrium that connects various new educational spaces.

The atrium contains comfortable study areas, with furniture scaled to the size of a room for study groups to settle into. The red carpet, together with the dark walnut wood, creates a subdued ambience throughout the public space, which is bordered by a 35 m long and 4-floor high bookcase. Opposite this, lecture halls protrude into the space. The study square spills over into the foyer, furnished with a service counter, a window with a view of the city, and an extensive outdoor terrace. Beyond the window, passageways running around the edge of the square's upper floor allow access into the lecture halls. Remnants from its previous construction, the sculptural concrete staircases cascading from the passageways once belonged on the building's exterior; now, they form part of the interior.

Excavating new space out of existing buildings is the ultimate architectural exercise. The characteristics of the original building were thoroughly researched. While it may seem contradictory, this knowledge was strategically used to germinate a design that creates a new atmosphere. Consistent with the pattern of the existing complex and by employing existing supports, the study centre presents itself as an open, unbound space. The exterior view of the center is equally imposing. Rising from the anonymous substructure, the articulated cube with an enormous window offers a view out and a look in.

The idea of superimposing a new roof over part of the existing complex is an intervention to enlighten the space. The roof itself, brings light in with its man-sized hollow rafters and glass roofing. The underlying 10 × 10 m grid structure – its maximum prefabricated size – gives the ceiling a clean look, along with the pattern of day-lit triangles. The roof creates a feeling of space, provides daylight, general lighting, and diffuses noise. Concrete beams – a gift from the 50 year old construction – are integrated to bear the load.

The Erasmus Medical Center houses separate entities of academic education, research, and patient care facilities together under one roof. They are connected by a wide communal corridor through the complex. 2 monumental staircases indicate the location of the interior entrance to the study center. The slim, white spirals presage the sea of space and light inside.

Credits and Data
Project title: Education Center Erasmus MC
Client: Erasmus MC Directie Huisvesting
Location: Rotterdam, the Netherlands
Design: 2006–2009
Completion: March 2013
Architect: KAAN Architects
Design team: Allard Assies, Luca Baialardo, Timo Cardol, Sebastian van Damme, Luuk Dietz, Paolo Faleschini, Raluca Firicel, Michael Geensen, Renata Gilio, Walter Hoogerwerf, Michiel van der Horst, Kees Kaan, Giuseppe Mazzaglia, Eric van Noord, Hannes Ochmann, Vincent Panhuysen, Antonia Reif, Dikkie Scipio, Shy Shavit, Koen van Tienen, Aldo Trim, Noëmi Vos
Project team: KAAN Architects (project management), J.P. Van Eesteren (main contractor), Aronsohn Raadgevende Ingenieurs (construction advisor), Royal Haskoning (technical installations), DGMR (building physics), Wolter en Dros (water installations), Kuijpers (electrical installations)
Project area: 34,000 m² (gross floor area)

p. 144: Photo taken before the renovation. The space used to be an uncovered courtyard. Photo courtesy of the architect. p. 145: Photo taken after renovation. The new atrium space spans 4-floor and is surrounded by bookcases. pp. 146-147: Close up view of the bookcase and the entrance to the room at the back. Photos on pp. 144–155 by Fernando Guerra.

第 144 页：翻修前的样子。这个空间曾经是一个没有遮蔽的庭院。第 145 页：翻修后的照片。新的中庭空间纵跨四层楼，被书架包围。第 146-147 页：书架及后房入口近景。

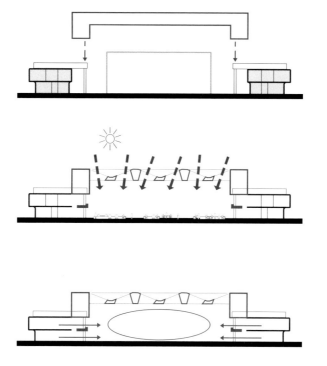

Concept diagram／概念图

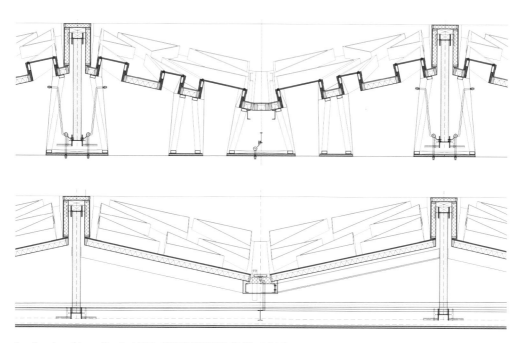

Detail section of the roof (scale: 1/100)／屋顶细部剖面图（比例：1/100）

This page: View from the corridor located west on the second floor.

本页：三楼西侧的走廊。

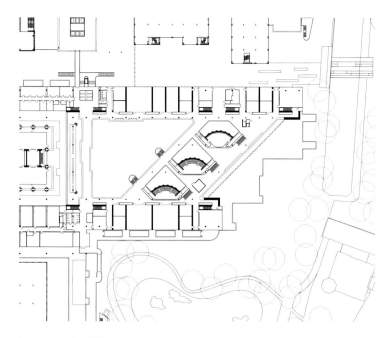

Second floor plan／三层平面图

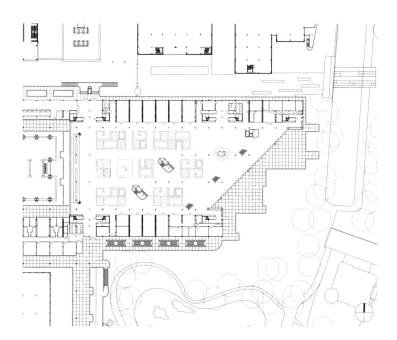

First floor plan (scale: 1/2,000)／二层平面图（比例：1/2,000）

医学教育研究中心位于伊拉姆斯大学的学术区，挑高的屋顶采用了大量的玻璃，营造出一种宽敞明亮的空间感。这里原本位于三楼。最初，铺设好的外院是为了引导使用者沿独具吸引力的路线穿越建筑，但这个连接空间从未真正发挥作用。现在，它已被改造为利用率更高的中庭，连接着各个新的教育空间。

中庭设有舒适的学习区，并配备了尺寸适合学习小组使用的家具。红地毯和深色胡桃木为整个公共空间营造了一种柔和的氛围。空间周围环绕着35米长、4层楼高的书架，对面是延伸至该空间的报告厅。学习广场与门厅相接，并设有服务台、城市景观窗以及宽阔的室外露台。景观窗外，人们可沿环绕广场上层的通道进入报告厅。雕塑般的混凝土楼梯是以前建筑的遗留物，自走廊螺旋而下。这部分曾经属于建筑的外部，现在，它们构成了室内的一部分。

从现有建筑中发掘新空间是一种终极建筑实践。虽然听起来矛盾，但通过深入研究原建筑所获取的知识是可以被策略性地用于设计中，并营造一种新氛围。研究中心利用了现有的支撑，呈现出与现有建筑模式一致的开放无约束的空间。中心的外观同样令人印象深刻，装配了大尺寸窗户的体块与底部结构铰接，提供了内外双向的景观视野。

在部分既有建筑上叠加新屋顶是一种激活空间的手法。通过大型的中空椽与玻璃屋面，自然光被引入室内空间中。下部的10米×10米的网格结构（最大预制尺寸）在天花板上呈现出三角形的日光图案，视觉效果干净整洁。屋顶在营造空间感的同时，也提供了自然采光、一般照明与噪声分散功能。具有50年历史的混凝土梁也被整合进来以承载负荷。

伊拉姆斯大学医学教育研究中心将学术教育、研究和患者护理设施等各独立单元整合在一起，通过宽阔的公共走廊相连。两个纪念性的楼梯指示了研究中心内部入口的位置，纤细的白色螺旋结构也预示着丰富的空间与充裕的光线。

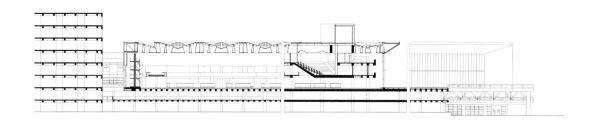

Long section (scale: 1/1,200)／纵向剖面图（比例：1/1,200）

Opposite, above: Interior view of the auditorium located between the second and third floor. Opposite, below: The newly built roof with its top light can be experienced in close proximity within this space.

对页，上：位于三楼和四楼之间的礼堂内景；对页，下：人们可以在这个空间内近距离体验带有顶灯的新建屋顶。

This page and opposite: Aerial view from the southeast. Photo by Aerophoto Schiphol - Marco van Middelkoop.

本页及对页：东南面鸟瞰图。

Frits van Dongen Architects and Planners, Koschuch Architects
Musis Sacrum
Arnhem 2015–2018

弗里茨·范·东恩建筑规划事务所，科舒奇建筑师事务所
圣珂兰音乐厅
阿纳姆 2015–2018

Since 1847, the concert halls of Musis Sacrum and its surroundings had taken on a range of different forms and positions in the park that reflected the short time needs of its operators, city officials, and residents. The project consists of a renovation of the existing monument and a new extension with long-term plans in mind. The extension features a large concert hall and supportive functions.

The design draws its origin from the core of the institution that defines Musis Sacrum: to perform and listen to music in excellent acoustical conditions within an attractive green context. The inspiration of the design reflects clearly the character of the Park, the Musis Sacrum Institute, and the identity of the Gelders Philharmonic Orchestra with its concert hall. No concessions are made regarding multi-functionality and acoustics, enabling the Musis Sacrum institute to be a home for all kinds of shows and events – ranging from symphony concerts to pop gigs and everything in between.

The extension is designed as an inviting and transparent pavilion respectfully complementing the historic building in its beautiful surroundings. The new large concert hall features a large glass window behind its stage and acts as a botanical backdrop. But, its unique quality is that the large glass window can also be opened for outdoor performances. Unlike its existing extension, the new multi-purpose hall is positioned as a separate volume in the park, thereby ensuring that our design gives the Musis Sacrum both the stage and space it deserves.

Site plan (scale: 1/6,000)／总平面图（比例：1/6,000）

自 1847 年以来，坐落于公园中的圣珂兰音乐厅及其周边设施以不同形式和位置回应着其经营者、城市官方及城市居民的短期需求。该项目作为一个长期计划，包含了整修现有纪念碑以及新的扩建项目，扩建的部分包含一个大型音乐厅及其配套设施。

设计源于圣珂兰的核心理念：在绿色自然景观与极佳声学条件下表演与聆听音乐。设计的灵感清晰地反映了公园的特点，以及圣珂兰机构、盖尔德爱乐乐团与音乐厅的身份特征。建筑师力求在多功能性和声学方面达到最佳效果，使圣珂兰音乐厅成为从交响乐到流行乐都适用的、能够举办各类演出与活动的场所。

扩建部分被设计成一个充满吸引力的透明"展亭"，以一种尊重的姿态在优美的环境中回应历史。新的大型音乐厅舞台后方有一面朝向公园的大玻璃窗，为演出提供了植物背景。它的独特之处在于，大玻璃窗也可以为户外演出打开。不同于原有扩建部分，建筑师将新的多功能厅设计成公园中的一个独立体量，以确保为圣珂兰音乐厅提供了应有的舞台与空间。

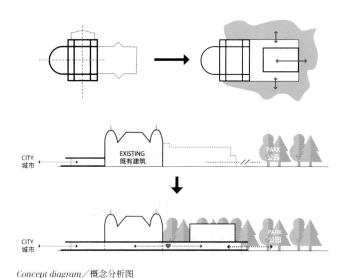

Concept diagram／概念分析图

p. 156: Former Musis Sacrum. The building was thoroughly renovated in 1995, prior to the completion of the new extension completed in 2018. Photo courtesy of De Gelderlander. p. 157: Southwest view of the new extension of Musis Sacrum from across the lake. Opposite, above: Exterior view from the west where one can see the boundary between the original 1874 building and its new extension. Opposite, below: Exterior view of the main hall from the south. Photos on pp. 156–165 by Bart van Hoek.

第 156 页：改造前的音乐厅。该建筑于 1995 年进行了全面翻新，2018 年完成新扩建。第 157 页：从湖对面看去，新扩建音乐厅的西南面外观。对页，上：西面外观，人们可以看到原始建筑 (1847 年) 和新扩建建筑之间的边界；对页，下：南侧主厅外观。

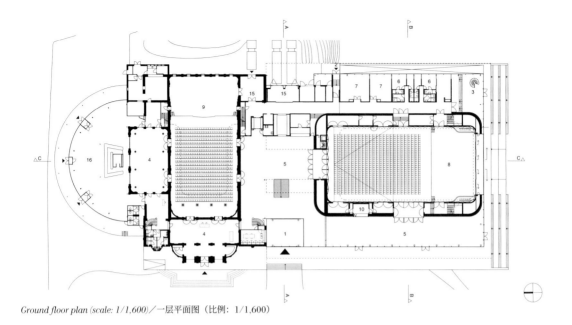

Ground floor plan (scale: 1/1,600)／一层平面图（比例：1/1,600）

1. Main entrance
2. Artist entrance
3. Artist foyer
4. Foyer
5. New foyer
6. Changing room
7. Office
8. Main hall
9. Classic hall
10. Bar
11. Kitchen
12. Wardrobe
13. Storage
14. Installations room
15. Loading area
16. Restaurant

1. 主入口
2. 艺术家入口
3. 艺术家休息室
4. 休息室
5. 新的休息室
6. 更衣室
7. 办公室
8. 主厅
9. 经典大厅
10. 酒吧
11. 厨房
12. 衣柜
13. 储藏室
14. 安装部
15. 装货区
16. 餐厅

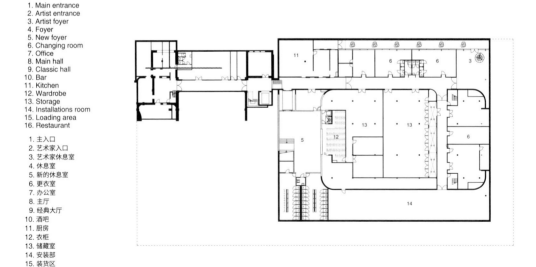

Basement floor plan／地下层平面图

Opposite, above: Foyer on the ground floor of the new extension. Opposite, below: Cloakroom counter on the basement floor.

对页，上：新扩建音乐厅一楼的门厅；对页，下：地下层的衣帽间寄存处。

Credits and Data
Project title: Musis Sacrum
Client: Gemeente Arnhem, Musis & Stadstheater Arnhem
Location: Arnhem, the Netherlands
Design: 2015
Completion: 2018
Architect: Frits van Dongen Architects and Planners, Koschuch Architects , Koschuch Architects
Design team: Ralph van Mameren, Elisabetta Bono, Rui Duarte, Hesh Fekry, Maikel Super, Casper de Heer, Klaas Sluijs, Olga Moreno
Project team: Theateradvies B.V. (theatre advisor), Peutz B.V. (acoustics and engineering advisor), Nelissen Ingenieursbureau (installations advisor), VandeLaar (structure engineer)
Project area: 5,950 m² (new), 3,700 m² (extension), 3,870 m² (existing)

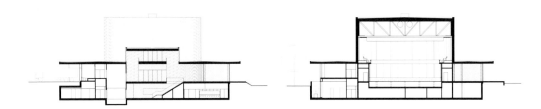

Section A／A剖面图

Section B／B剖面图

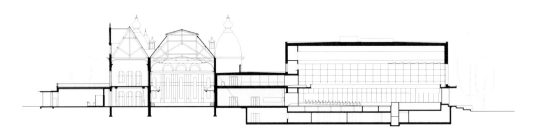

Section C (scale: 1/1,800)／C剖面图（比例：1/1,800）

pp. 162–163: Main hall. This page: The classic hall of the original building. Opposite, above: Performer waiting lounge. Opposite, below: Waiting area of the foyer on the ground floor.

第 162-163 页：主厅。本页：原建筑的经典大厅。对页，上：表演者等候厅；对页，下：一楼门厅的等候区。

Happel Cornelisse Verhoeven
Museum De Lakenhal
Leiden 2014–2019

哈佩尔-科尼利斯-韦尔霍芬建筑师事务所
莱克纳尔博物馆
莱顿 2014–2019

pp. 166–167: North facade of the new extension. This page, above: View of the courtyard during the construction. This page, below: View of the gallery during the construction. Photos on pp. 122–128 by Karin Borghouts.

第166-167页：新扩建的北立面。本页，上：施工期间的庭院；本页，下：施工期间的画廊。

Laecken-Halle (1664)
莱肯哈勒 (1664)

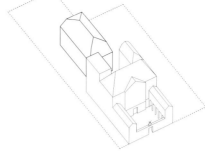

Harteveltzaal (1890)
哈特维尔扎尔 (1890)

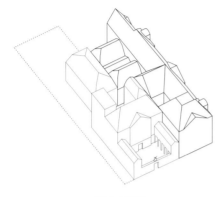

Papevleugel (1921)
帕夫洛韦格 (1921)

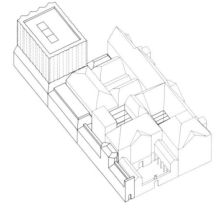

Van Steijn Gebouw (2019)
范·史戴恩·格布 (2019)

Construction diagram／建造分析图

Museum De Lakenhal has been the municipal museum of art, arts and crafts, and the history of the city of Leiden since 1874. It is located in the monumental "Laecken-Halle" from 1641 by city architect Arent van 's-Gravesande, built for the approval of cloth fabrics. Today, the museum ensemble consists of buildings from the 17th, 19th, 20th and 21st centuries, each with its own characteristics. The point of departure for its restoration and expansion was to create a balance among these layers of time, according to the principle of "unity in diversity" where the orientation and experience of art are spatially improved and supported.

The restorative work for De Lakenhal was used to reinforce its 17th century individuality. Using the same material palette and furniture design, the various buildings were subsequently matched; while the reception, orientation, and art experience have been spatially improved. One of the most important interventions, in the heart of the ensemble, is the clearance of the former "Achterplaets" (backyard). This exterior space is fitted with a glass roof and forms the central orientation space of the museum. From here, the various building parts are visible and accessible. Traces of almost 375 years of building history have not been erased and are left to be seen.

In addition to the restoration of the monumental building complex, a new building volume is added on the west and extends telescopically between the Oude Singel and the Lammermarkt. The ground floor houses 2 exhibition halls, a museum restaurant, and an indoor loading and unloading zone. On the Lammermarkt side, the new building presents itself as a "shouldered" figure in which offices, studios and the library are housed. Like the "Laecken-Halle" and its later additions, the new building presents itself as a recognizable one-piece architectural unit. It is a new brick offspring of the "family".

Site plan (scale: 1/4,000) ／总平面图（比例：1/4,000）

Credits and Data

Project title: Museum De Lakenhal Client: Municipality of Leiden
Location: Leiden, the Netherlands Design: December 2013
Completion: May 2019

Architect: Happel Cornelisse Verhoeven (main architect), Julian Harrap Architects (restoration)

Project team: IBB Kondor, Koninklijke Woudenberg, Brandwacht en Meijer (contractor), Van Rossum (structural engineer), Arup (technical engineer), LPB Sight (acoustics), Happel Cornelisse Verhoeven (interior architect), Beersnielsen lichtontwerpers (light design), Karin Polder (graphic design), Aleksandra Gaca, Iemke van Dijk, Ankie Stoutjesdijk, Hansje van Halem, Studio Maarten Kolk & Guus Kusters (art integration)

Project area: 8,700 m²

Project estimate: 16,123,000 euro (without tax)

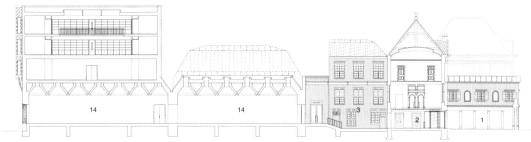

Long section (scale: 1/600) / 纵向剖面图（比例：1/600）

1. Forecourt
2. Vestibule
3. Achterplaats
14. New exhibition galleries

1. 前院
2. 前厅
3. 后院
14. 新展览画廊

p. 170: South facade is restored to its original form. p.171: Museum cafe fitted with an entrance gate from 1660. Opposite: Interior view of the museum cafe. This page: A new glass roof is placed over the courtyard.

第170页：恢复原状的南立面。第171页：1660年当时的门被嵌入博物馆咖啡厅。对页：博物馆咖啡厅内部。本页：位于庭院上方的新玻璃屋顶。

This page, clockwise from top left: Fine Painters Galleries. University room on the third floor. Atelier. Gallery of the extension building on the north. Entrance. Auditorium.

本页，自左上顺时针方向分别为：美术家画廊，四楼大学室，工作室，北部扩建大楼的画廊，入口，礼堂。

莱顿莱克纳尔博物馆自1874年以来一直是艺术、手工艺品博物馆以及莱顿市的历史代表设施。它位于城市建筑师阿伦特·范·格雷夫桑德于1641年建造的、具有纪念意义的"莱肯哈勒"（建筑名）中。如今，博物馆整体由17、19、20和21世纪各具特色的建筑组成。其修缮和扩建根据"多样性统一"的原则，在这些时间层次中建立平衡，从而使艺术体验和方向性在空间上得到改善和支撑。

为了增强其17世纪的个性，莱克纳尔博物馆的修复工作使用相同的材料组合和家具设计，各个建筑物随之匹配。同时，接待、导览和艺术体验在空间上也得到了改善。在建筑群中心，最重要的干预措施之一是修整了留下的"Achterplaets"（后院）。外部空间配有玻璃屋顶，这在博物馆的中心形成了一个具有向心力的空间。在这里，建筑各个部分都是可见并且可达的。已有近375年历史的建筑痕迹尚未消除，尚待观赏。

除了对具有纪念性的建筑群进行修复外，西侧还增加了一个新的建筑体，并在奥德·辛格尔街和兰马多街之间进行了延伸。一楼设有两个展览厅，一间博物馆餐厅和一个室内装卸区。在兰多马街一侧，新建筑以一个"有担当"的形象呈现，办公室、工作室和图书馆都置于其中。就像"莱肯哈勒"及后来增建的建筑单元一样，新建筑以一个有辨识度的整体建筑单元的形式呈现。它是这个"家庭"的新成员。

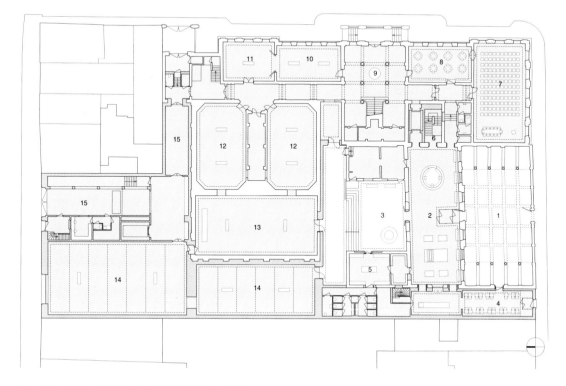

Ground floor plan (scale: 1/600)／一层平面图（比例：1/600）

NL Architects, XVW Architecture
Kleiburg De Flat
Amsterdam-Zuidoost –2016

NL建筑师事务所，XVW建筑师事务所
克莱堡公寓
阿姆斯特丹，东南区 –2016

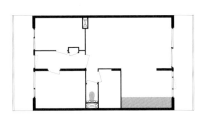

Unit Type A Plan, H#735 ／单元类型A平面图，735号室

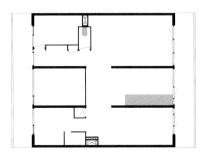

Unit Type B Plan, H#803 ／单元类型B平面图，803号室

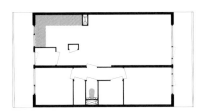

Unit Type A Plan, H#132 ／单元类型A平面图，132号室

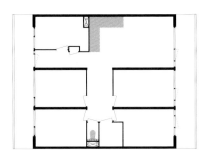

Unit Type B Plan, H#526 ／单元类型B平面图，526号室

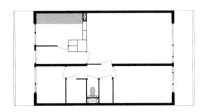

Unit Type A Plan, Original (scale: 1/250)
单元类型A平面图，改造前（比例：1/250）

Unit Type B Plan, Original (scale: 1/250)
单元类型B平面图，改造前（比例：1/250）

Kleiburg is one of the biggest apartment buildings in the Netherlands: a bend slab with 500 apartments, 400 m long, and 10 + 1 stories high. Kleiburg is located in the Bijlmermeer, a CIAM inspired residential expansion of Amsterdam designed in the 60s. An urban renewal operation started in the mid-90s, and the characteristic honeycomb slabs were replaced by mostly suburban substance – by "normality".

Kleiburg was the only remaining fragment in a more or less original state. However, the Housing Corporation Rochdale calculated that a thorough renovation would be too costly, and demolition seemed to be the only option. But, there was fierce resistance and eventually, Kleiburg was offered at 1 euro in the attempt to catalyze alternative, economically viable plans. Consortium deFlat was chosen with its proposal to turn Kleiburg into a Klusflat. "Klussen" translates as to do it yourself. The idea was to renovate the main structure (elevators, galleries, and installations) but leave the apartments unfinished (no kitchen, no shower, no heating, no rooms). Its future residents could buy the shell at an extremely low price and then renovate it entirely according to their wishes. Owning an ideal home suddenly came within reach. The smart renovation revealed its latent beauty and established a successful embedding in the park that surrounds it. Kleiburg was miraculously brought back to life – a collective effort of more than 500 people.

Credits and Data
Project title: Kleiburg De Flat – Residential Revamp of a Modern Monument
Client / Concept developer: Consortium De Flat – KondorWessels Vastgoed, Hendriks CPO, Vireo Vastgoed, Hollands Licht (Martijn Blom)
Location: Amsterdam-Zuidoost, the Netherlands
Design: 2011
Completion: 2016
Architect: NL Architects, XVW Architecture
Design team: (NL Architects) Pieter Bannnenberg, Walter van Dijk, Kamiel Klaasse, Guus Peters, Iwan Hameleers, Giulia Pastore, Fouad Addou, Matthew Davis, Paul Ducom, Soo Kyung Chun, Adrian Mans, Paulo Dos Sousa, Carmen Valtierra de Luis, (XVW Architecture) Xander Vermeulen Windsant, Patrick Wozniak, Bartjan van Slot
Project team: Van Rossum Raadgevende Ingenieurs Amsterdam B.V., Schreuder Groep (building physics), HOMIJ Technische Installaties B.V., (installations), KondorWessels Amsterdam B.V. (contractor)
Project area: 65,600 m² (gross floor area)

p. 177: Photos depicting the "do-it-yourself" character of the apartment interiors. From the top, left-to-right: Photo 1 by Stijn Poelstra. Photos 2 and 7 by Marcel van der Burg. Photos 3–6 by Stijn Brakkee. Opposite: View of the conserved original facade. Photo by Stijn Poelstra.

第 177 页：照片描绘了公寓室内住户可自己设计改装的"do-it-yourself"构想。对页：保留下来的原有立面。

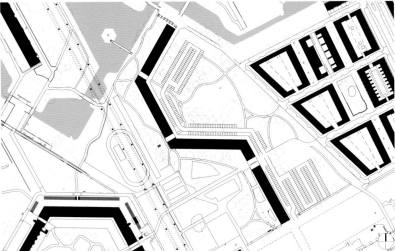

Site plan (scale: 1/5,000)／总平面图（比例：1/5,000）

This page: Aerial view of Kleiburg. Photo taken during the 1960s. Photo courtesy of Stadsarchief Amsterdam. Opposite: Community spaces like tennis courts are provided around the neighborhood. Photo by Marcel van der Burg. p. 182, above: View from the ground floor. p. 182, below: Community garden. Photos by Stijn Brakkee.

本页：克莱堡鸟瞰图。照片摄于 1960 年代。对页：社区周围设有网球场等社区空间。第 182 页，上：一层视野；第 182 页，下：社区花园。

克莱堡是荷兰最大的集合住宅建筑之一，这个 400 米长、10+1 层高的折线板式建筑中共有 500 套公寓。克莱堡位于比莫米尔街区，这是一个于 1960 年代设计的，受国际现代建筑协会（CIAM）启发的阿姆斯特丹住宅扩建项目。1990 年代中期开始的城市更新行动中，标志性的蜂窝状板楼大多被郊区的"规范"建筑所取代。

克莱堡是仅剩的一处部分保有原始状态的建筑。然而，罗奇代尔住房公司的计算显示，彻底翻新的成本过高，拆除似乎是唯一的选择。在激烈的抵制下，克莱堡最终以一欧元的低价出售，试图促成在经济上可行的转型计划。最后中选的方案来自 deFlat 财团，他们提议将克莱堡转变为一个"克鲁斯法特"。"Klussen"翻译过来是"自己动手"的意思。该想法仅翻新主体结构（电梯、画廊和装置），而将公寓保留在一种未完成的状态下（没有厨房、淋浴、暖气、甚至房间）。未来的住户可以用极低的价格购入公寓的外壳，并完全按照自己的意愿进行翻新。

拥有一个理想的家突然间变得触手可及。这种巧妙的改造展现了克莱堡的潜在之美，并成功地将其嵌入周围的公园中。克莱堡奇迹般地恢复了生机，这是 500 多人共同努力的结果。

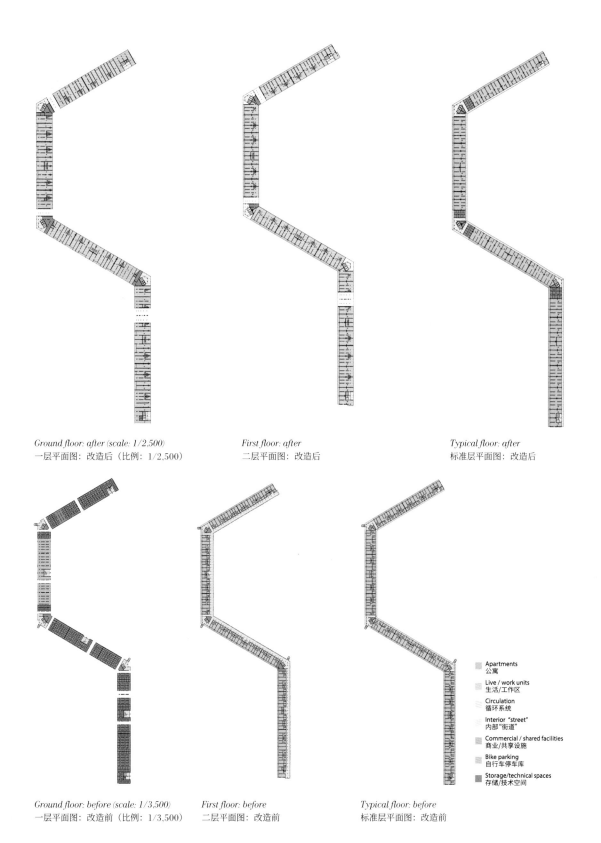

Ground floor: after (scale: 1/2,500)
一层平面图：改造后（比例：1/2,500）

First floor: after
二层平面图：改造后

Typical floor: after
标准层平面图：改造后

Ground floor: before (scale: 1/3,500)
一层平面图：改造前（比例：1/3,500）

First floor: before
二层平面图：改造前

Typical floor: before
标准层平面图：改造前

Apartments
公寓

Live / work units
生活/工作区

Circulation
循环系统

Interior "street"
内部"街道"

Commercial / shared facilities
商业/共享设施

Bike parking
自行车停车库

Storage/technical spaces
存储/技术空间

Essay:
After the NAi: Grand Projects in the Netherlands
Sergio M. Figueiredo

论文:
NAi之后：荷兰的宏大项目
塞尔吉奥·M.菲格雷多

Between 1993 and 2013, the heart of Dutch architecture culture stood in Rotterdam's Museumpark. At the head of the park, the Netherlands Architecture Institute (NAi) was housed in an intriguing building composed of discrete volumes of varied material expression – a glistening glass tower, a solid brick-covered cube, and an elevated, sweeping, metal-clad curve – expressing the multitude of activities conducted inside: archives, research, exhibitions, and public discussion. While never fully recognized, this building already indicated how Dutch architecture would become renowned the world over, that is, as ambitious architectural statements both revealing and commenting on their very program and surrounding social conditions. An architecture that was best understood from afar was easily translated and disseminated through imagery. Together with the Dutch government's progressive policies on architecture, the NAi was the most prominent force in shaping Dutch architecture culture, relentlessly supporting the production and appreciation of quality architecture in the Netherlands. Its message was clear: architecture was not limited to the construction of walls and roofs, windows and doors, but was also, fundamentally, a priceless cultural endeavor.

On 1st January 2013, however, as the NAi ceased to exist, Dutch architecture culture – and with it, Dutch architecture – was irrevocably changed. In the NAi's place, the Het Nieuwe Instituut (HNI) was established, a new institution dedicated to all "creative industries" resulting from the NAi's forced merger with the Netherlands Institute for Design and Fashion (Premsela) and the industry institute for electronic culture (Virtueel Platform).[1] The NAi's closure, however, was but an indication of how the famed Dutch architectural infrastructure was unceremoniously dismantled as the Dutch government pursued a commercialization of culture, primarily driven by an ideological – and almost dogmatic – belief in the invisible hand of the free market. Specifically, in its commitment to a market-driven approach to culture, the Dutch government determined that architecture was no longer an independent cultural sector that required its own support infrastructure, but rather a component of the new creative industries sector, which had been identified as a strategic area for the country's economy.[2] As a result, not only was the NAi merged out of existence, but smaller architectural organizations that could not demonstrate immediate economic return – such as Archiprix, Europan (Netherlands), Architektuur Lokaal, the International Architecture Biennale Rotterdam (IABR), and even the Berlage Institute – were completely defunded.

Beyond a visible demotion of architecture's public standing, its changed status also indicated a crucial ideological shift in the very understanding of architecture in the Netherlands. Effectively, as a creative industry, architecture's cultural ambitions were supplanted by economic returns, as "attention to artistic autonomy [was] displaced by user demands and marketability" while "results were often meant as economic results."[3] With this paradigm shift, the previously blurred boundaries between architecture theory and practice were clarified and the two domains effectively severed. The strategies of engagement that had been built up over the previous twenty years – through which the development of Dutch architecture was directly

993年至2013年间,荷兰建筑文化的心脏始终位于鹿特丹的博物馆公园。公园一端的荷兰建筑协会(NAi)被包裹在一栋有趣的建筑中,它由性质各异的独立体块组合而成,包括闪耀的玻璃塔、覆盖砖块的立方体、飞扬的金属曲线。这象征着建筑内部的丰富活动:记录、研究、展览和公共讨论。但人们没有意识到,它还暗示着荷兰建筑终将享誉世界,即通过大胆的建筑表现,对项目本身及其所处的社会环境进行揭露和评判。由于远观更能理解设计意图,这的建筑形象被广泛转译、传播;再加上荷兰政府积极的建筑政策,NAi便成了塑造荷兰建筑文化最重要的一支力量,提升着荷兰建筑的质量与声誉。它明确传达出这样的信息:建筑不局限于建造墙壁、屋顶和门窗,从根本上说,它也是一种无价的文化结晶。

然而,2013年1月1日,随着NAi的解散,荷兰建筑文化以及与之共存的荷兰建筑都发生了不可逆转的变化。取而代之的新研究所(Het Nieuwe Instituut, HNI)是NAi、荷兰设计与时尚协会(Premsela)、电子文化产业协会(Virtueel Platform)被迫合并成立的新机构,负责所有的"创意产业"[1]。事实上,NAi的关闭更意味着,即使享有盛誉,荷兰建筑的根基仍被破坏,而荷兰政府还在有意识地,甚至几乎教条化地,信任自由市场的无形之手,追求文化的商业化。具体而言,荷兰政府决定把文化交给市场,不再将建筑视为独立的、需要专项支持的文化部门,而是将其列入国家经济的战略性增长领域——创意产业部门[2]。最终,不义NAi被合并,那些无法快速产生经济效益的小型建筑团体,例如阿奇普里克斯基金会、欧罗巴(荷兰)基金会、洛卡功建筑基金会、鹿特丹国际建筑双年展(IABR)、贝尔拉格学院,也都完全失去了资金。

不只是建筑的公共地位明显下降,这种变化还表明任何二,人们对建筑的理解发生了天翻地覆的转变。作为创意产业,建筑的文化野心被经济效益取代,因为"对艺术自主性的关注被用户需求和市场价值取代",而"结果往往意味着经济成果"[3]。伴随这种模式的转变,建筑理论和实践之间原本模糊的界限变得清晰,二者被彻底分离。前二十年建立起来的政治参与策略直接或间接影响了荷兰建筑的发展,但与现在几乎完全抽身的情形,形成了鲜明的对比。

荷兰中央政府难辞其咎:他们不仅忽视了引导建筑讨论和欣赏的责任,还放弃了建筑生产的责任。尽管这在NAi消亡之前就已出现,甚至可以追溯到荷兰建筑政策的制定,但在2012年极度恶化。中央政府彻底抛开积极严苛、雄心勃勃的建筑委托方的角色,主张民间发起、地方管理。由此造成了两大影响,即中央政府将不再为建筑委托制定有效、细致的标准,也不会再有全面、调和的建筑政策。最终,建筑的文化愿景和社会意图越来越多地交由民间和地方机构实现,但如果没有中央政府的金融资源和政策支持,可想而知这些不会多么远大。

实际上,中央政府的新自由主义做法已经严重影响到了建筑的公共鉴赏和生产。简而言之,这意味着市场化程度较低的建筑方法不再有任何结构性的支持,而缺乏了NAi和类似机构的持续关注与讨论,荷兰建筑会从公众话题中隐身,以至于被遗忘。在过去,是这些机构和公共委托给新人建筑师提供了实验的机会,让他们的声音逐渐变得有力,同时促进大众理解建筑的意图和雄心;但新的市场驱动形式就意味着,只有得到资助,方案才会被落实,甚至仅仅被提出和讨论。

没有NAi(和留到最后的支撑基础),荷兰建筑就失去了一项重要的软实力,无法再向公众传达建筑独特的附加价值。NAi存在时,对建筑的理解、展示和讨论不仅属

and indirectly shaped – were now contrasted with an almost complete withdrawal.

Beyond neglecting its responsibility to direct the discussion and appreciation of architecture, the Dutch central government also abdicated from its responsibilities in architectural production. Although this process had been initiated well before the NAi's demise – and can be traced in the development of the Dutch architecture policies – it came to its head in 2012, as the government completely retreated from its role of active and demanding architectural client with high ambitions, favoring instead private initiative and local governance. Such retreat had 2 main implications, namely that the central government would no longer set the standard for effective and deliberate architectural commissioning, but also that there would no longer be a comprehensive, coordinated architecture policy. As a result, private parties and local institutions became increasingly responsible also for architecture's cultural and social intentions, which, without the central government's financial resources and institutional support, were predictably less ambitious.

In practical terms, the central government's neo-liberal approach had a significant impact on the public appreciation and the production of architecture in the Netherlands. In short, it meant that there was no longer any structural support for less marketable approaches to architecture, but also that without constant attention and discussion from the NAi and similar institutions, in the Netherlands architecture began to recede from public debate, almost to the point of oblivion. If in the past, these institutions and public commissioning had allowed for emerging architects to experiment and mature their own architectural voice while also supporting a greater appreciation for architecture's intentions and ambitions from the general public, the new market-driven approach to the discipline meant that only those ideas that could be funded would ever be supported or even presented and discussed.

Without the NAi [and remaining support infrastructure], Dutch architecture lost a crucial soft-power apparatus that had been geared towards conveying to the public the unique added value of architecture. With it, architecture had been consistently conceived, presented and discussed as something that – beyond technical, functional and pragmatic concerns – also comprised of an intrinsic cultural, societal, intellectual, and even artistic dimensions. Without it, however, architecture's dissemination was visibly hollowed out of any theoretical or critical ambitions and, eventually, of its cultural standing. Furthermore, as the economic crisis hit Dutch offices particularly hard – with, at its peak, a contraction of over 80% of new commissions and the dismissal of over a third of its workforce – to which the most widely contemplated solution was a more aggressive entrepreneurial attitude, it became increasingly clear that the time of audacious experiments and provocative expressions had come to an end.[4] In response to the newly empowered market forces, Dutch architecture ceased to be challenging to become accommodating. In the eye of the [economic] storm, and lacking the institutional resources to support its ambitions, architecture in the Netherlands went from a fresh and distinctive avant-garde to an almost banal

文本、功能、实用的层面,还包括内在的文化、社会、知识乃至艺术。然而如果没有它,建筑理论或批判显然就会趋于空洞,最终危及建筑的文化立场。此外,经济危机重创了荷兰的建筑事务所,极端时项目委托缩减80%以上,超过三分之一的员工被解雇,万全之策是采取更加大刀阔斧的措施;同时越来越明显的是,大胆实验和激进表达的时代已经结束了[4]。面对新近被赋权的市场,原先富于挑战的荷兰建筑变得温和;而由于没有政策支持,本就在(经济)风暴中心的荷兰建筑更是从别具一格的先锋,转变为平庸无奇的后卫。

在过去的十年里,探索建筑极限越来越难,荷兰建筑只有作出让步和适应。在"后超级荷兰"时期(或所谓的"派寸之后"),荷兰建筑逐渐缺失的,不仅是它的独特和激进,还有批判和野心,充满寓意的符号被直白的后数字拼贴画取代[5]。荷兰建筑的价值不再在于它包含的信息和图像,而变成了它提供的材料和体验。当建筑没有能力表达更宏大的意图、也无法反映社会时,就不得不再次关注直白的空间体验。因此,对荷兰建筑的定义不再是醒目的形象、大胆的理念或尖锐的话语,它的特征越来越倾向于更温和的设计;然而,一旦真面目被揭开,它还会以不那么强势却依旧引人注目的方式出现在我们眼前。

要清晰反映荷兰建筑的这一新阶段,恐怕没有比科恩·范·维尔森的布雷达汽车站与火车站(见220-231页)更适合的(宏大)项目了。这个姿态谦和、默默无闻的庞然大物,一边模仿、融入所处的环境,一边将错综复杂的组合隐藏在巨大的屋顶之下。多种砖材的巧妙组合控制住了建筑的体量,只有棱角分明的几何外观凸显出整体的宏大。而这番质朴的外表下,却有一个丰富的内在——温暖的材料、舒适的小空间,意想不到的小角落,都缓和了交通枢纽的沉闷,形成一种独特迷人的氛围。精心设计的框架式入口突出了日常的通勤活动,也在视觉上连接起了列车站台、公交车站和停车区。

在这里,建筑的内敛性被推至极致,实践没有基于任何理论化的、肤浅的宣言。这个公共交通枢纽不是在对时代问题或社会愿景发表什么宏论,它只是一个简单的、直截了当的巨大建筑,通过使用和体验来展示自己。因此,选用合适材料,提高空间质量,便是这一建筑设计的主要特征。

这种室内密集装饰与外观低调呈现的对比,也体现在21世纪最宏大的项目,阿姆斯特丹的荷兰国家博物馆(见232-245页)上。总部位于塞维利亚的克鲁兹·奥尔蒂斯建筑师事务所,基于清理建筑庭院的简单想法,将1960年代临时设置的展厅移除。整个翻新工程考虑周全,最终优雅融入了皮埃尔·克伊珀斯设计的这栋19世纪建筑之中。清理后的庭院和创造出的中庭(两个庭院由地下通道连接,不仅为正在老去的建筑提供了新的方向和秩序,还成了项目标志性的建筑元素。

宽敞的新中庭既可容纳所有常规的博物馆服务,又是通往上方展厅的入口。光线透过巨大的玻璃拱顶和雕塑般的笼状吊灯,笼罩着整个中庭。地面和墙面完全覆盖葡萄牙石灰石,进一步突出中央中庭的轻盈,带来视觉上的连贯性。丰富的材料包裹着参观者的感官,散发出宁静的气息。一系列单体入口标志着空间的转折,同时将尺度缩小至人体尺度。除了中央通道两侧的巨大窗户,这些改造很少表现在建筑外部,因为它更注重创造有特色的空间体验,而非向外展现这些意图。

同样注重空间体验的,还有麦肯诺建筑师事务所设计、新近竣工的代尔夫特市政厅和火车站(见208-219页)。建筑集行政服务、市政办公、火车站于一体,在内敛的

rear-garde.

In the past 10 years, as it became more difficult to take risks and explore architecture's edges, Dutch architecture simply retreated and adapted. In a post-SuperDutch period [or "after the party" as some have dubbed it], Dutch architecture became less iconic and less provocative, but also less critical and less ambitious, as rhetoric-filled diagrams were replaced by mood-conveying post-digital collages.[5] Dutch architecture's value was no longer to be found in its message and image, but rather in its materiality and experience. Without the ability to articulate grander intentions or societal commentaries, it was necessary to focus, once again, on the immediacy of spatial experience. Therefore, while no longer defined by strong images, challenging ideas or bold discourse, Dutch architecture became increasingly characterized by more unassuming designs that unravel before our eyes in less forceful, yet equally striking ways.

No other [grand] project expresses this new stage of Dutch architecture as clearly as Koen van Velsen's Public Transport Terminal Breda (See pp. 220–231). This gentle giant, unassuming almost to the point of anonymity, mimics and blends with its context while hiding an ingenious programmatic conflation under its enormous roof structure. Hints of monumentality are only asserted by the building's robust geometric expression, as its enormous size is carefully controlled and subdued by its careful materialization in an assortment of bricks. Its unpretentious exterior is met with a richly textured interior – where warm materials, a profusion of smaller spaces, and otherwise unexpected nooks soften the terminal's institutional character and set a distinctive inviting tone. Strategically placed voids frame that celebrates the everyday life of commuters, while creating visual connections across train platforms, bus stops, and car parking decks.

Here, the introversion of architecture is taken to an almost sublime extreme, in which the practice of architecture is devoid of any theorization or facile proclamation. Without any sweeping statements about contemporary conditions or social visions, this public transport terminal is a simple, no-nonsense, architectural monument, that reveals itself through use and experience, where spatial quality – enhanced by a precise materiality – becomes the design's dominant architectural feature.

Such contrast between interior material intensity and exterior subtleness of expression was already observed in the grandest of all projects in the Netherlands in the 21st century, the Rijksmuseum in Amsterdam (See pp. 232–245). With the simple idea of clearing the building's courtyards of their 1960s ad-hoc infills of additional galleries, the Seville-based office Cruz y Ortiz was able to create a thoughtful renovation that blended gracefully with P.J.H. Cuypers original 19th century building. The clearing of the courtyards and the creation of the central atrium [connected underneath the central passageway] not only provided a new orientation and organization to the aging building, but also became the project's most notable architectural element.

Accommodating all the usual museum services and acting as an entry point to the galleries above, the new ample atrium, is bathed in light, filtered through enormous glass vaults and a sculptural set of cage-like chandeliers. The central atrium's

外观之下，将大部分建筑效果藏于室内。从外部来看，建筑呈现为一个紧凑、不对称、有棱有角且带有几处深邃切口的体块，完全被玻璃板覆盖（其中一些可看出是传统的窗户）；内部却颇为开阔，独特的拱形天花板上，还用当地特有的代尔夫特蓝瓷拼贴出了昔日的城市地图。从车站到市政厅，你总可以沿着墙壁和立柱，找到代尔夫特蓝瓷在当代的重新诠释。这样一个建筑，外观即使不能说彻底融入环境，也普通到仿佛随处可见；然而正是它致力于创造独特的空间体验，隐喻着代尔夫特的过去、现在和未来。

此外，埃克特·胡格斯塔德设计的乌得勒支自行车停车场也有类似的情况（见200-207页）。这个规模可观的地下设施几乎没有任何外部表现，特定的建筑细节却让它富有效率和活力。这些细节没有以图像呈现，而是直接影响着体验。虽然在地面上只能看到一个巨大的白色天幕，以及垂直通到停车场的立柱，地下空间的丰富性并未显露，但通过精心设计的开口、视野不广的玻璃窗，仍能从外面窥探三层停车场的内部。停车场划分自行车道和人行道，以简洁的涂色进行区别。整个设计的最大亮点是光滑如雕塑般的混凝土立柱穿过挑空空间，支撑着上方天幕，为建筑创造出独特的形象。

总之，这一自行车存放处的特色在于其光滑的质感、温和的结构、大量的曲线；这些让我们意识到，空间体验已在主导荷兰建筑的走向。内敛谦逊、建构直接体验的建筑，正通过实际使用而非表象来展现自我。

过去的十年里，尤其是自2013年NAi关闭以来，荷兰的建筑已经彻底改变。随着荷兰中央政府放弃最后这一直接和间接影响建筑发展的工具，荷兰建筑无疑失去了光芒。新的市场体系中，没有强有力的制度支持，荷兰建筑的批判性大大降低，甚至完全消失。但也由于缺乏批判和理论的热情，荷兰建筑找到了新的目标——挖掘人们的审美情趣，在对真实物质和经验价值的追求中，回归建筑的基础。尽管荷兰建筑不再有批判力，但正如比利时著名评论家吉尔特·贝克特所说，建筑无论是作为某种有冲击力的形象，还是带来同样令人难忘的体验，都"必须不断破坏和再造（它的）社会环境，才能保护、接纳这个现实"。[6]如今，荷兰建筑的文化核心不再那么显而易见。它隐藏在众目睽睽之下，沉睡在每一个调动我们感官、丰富平淡日常的建筑碎片之中，唯独会展现给那些知道去何处寻找的人。

参考文献：
1. 了解更多关于NAi被强行合并的信息，可查阅塞尔吉奥·M.菲格雷多的《谁需要新的机构？告别NAi》，2013年8月20日，http://www.domusweb.it/en/op-ed/2013/01/17/who-asked-for-a-new-institute-afarewell-to-the-nai.html.
2. Piet Vollaard, Meer Met Minder, ArchiNed, 2011年6月10日，http://www.archined.nl/nieuws/2011/juni/meer-met-minder/.
3. NIADEC, 2013-2016"创意必要"政策，荷兰建筑学院，Design En e-Cultuur, 2012年2月, 8-9.
4. Samir Bantal等, 荷兰建筑"平静前的风暴"：2009-2010年度报告 [M] 鹿特丹：NAi出版社, 2010, 7.
5. OASE, OASE #67:派对之后：荷兰建筑2005 [M]. 鹿特丹：NAi出版社, 2005.
6. 吉尔特·贝克特与克里斯托夫·范·格鲁维合著，《扎根现实：建筑写作》中由吉尔特·贝克特撰写的"建筑无影"，根特：WZW editions and productions, 2011, 39.

lightness and visual coherence are further accentuated by its engulfing cladding – both on the floors and walls – in Portuguese limestone, providing a rich material expression that envelops visitors' senses and radiates serenity. Completing the composition, monolithic gateways establish a spatial threshold and introduce a form of human scale. Beyond the enormous windows flanking the central passageway, little of this intervention is expressed on the building's exterior, as it is more focused on creating a compelling spatial experience than on projecting outwards these intentions.

A similar focus on spatial experience can be found in the recently completed Delft City Hall and Train Station (See pp. 208–219), designed by Mecanoo. Combining a train station with city services and municipal offices, the building's exterior expression is subdued, leaving most of its architectural effect to be created by its interior. Therefore, while the building's exterior is presented as a compact, asymmetrical, angular volume with several deep cuts, entirely covered in glazed panels [some of which with a detail alluding to a traditional window type], the interior is open and generous, marked by its unique vaulted ceiling depicting a historical map of the city in traditional Delft Blue. A contemporary re-interpretation of Delft Blue tiles is also found along the walls and columns of both station and city hall. Ultimately – while not quite blending in – the building's exterior is rather generic and could find a home almost anywhere, its interior, however, aims to create a unique spatial experience constantly marked by allusions to Delft's past, present and future.

Also, a similar condition is experienced in Ector Hoogstad's design for Utrecht Bicycle Parking (See pp. 200–207). Almost without any exterior expression, the utilitarian character of this [still sizable] underground piece of transport infrastructure is enlivened by specific architectural details that operate in the realm of the immediacy of experience rather than mediated imageability. As an underground structure, its only visible expression above ground is articulated by a large, white canopy, whose columns stretch deep into the bicycle storage. This canopy, however, does not reveal the richness of the spaces created below, as these are dramatically revealed through strategically placed voids providing select vistas across the 3 levels of the bicycle storage. While providing less expansive vistas, glazed openings reveal and frame the bicycles within. Orientation and orchestration of cyclists and pedestrians' movements are both facilitated and indicated in the clear, yet subdued, color accent of surfaces and pavements. The design finds its culmination in the smooth, sculptural concrete columns that punctuate the main void, supporting the canopy above and creating a distinct image for this project.

Overall, this bicycle storage is defined by its soothing materiality, a placid tectonic expression with a few sweeping curves that remind us how architecture's spatial experience has taken control in Dutch architecture. Introversion and modesty are instrumentalized as architectural tropes to construct the immediacy of experience and allow architecture to be revealed through use rather than image.

In the past ten years, but particularly since 2013 when the NAi was closed, architecture has utterly changed in the Netherlands. As the Dutch central government abdicated from its last instruments for directly and indirectly shaping architectural development, Dutch architecture has certainly lost its shine. Absent strong institutional support, and within its new market-system, the possibilities for Dutch architectural to be critical were greatly reduced, if not altogether abolished. Devoid of critical and theoretical ambitions, Dutch architecture found a new purpose in exploring a range of aesthetic emotions and sensations, reverting to the architectural grounding found in the honesty of materials and the value of experience. With this shift, architecture in the Netherlands became devoid of its critical shadow, but as Geert Bekaert, the famed Belgian critic, once stated, architecture "must continually destroy and recreate [its] milieu so that reality can be secured and inhabited," be it in the form of strong images or of equally remarkable experiences. Today, the heart of Dutch architecture culture is no longer so easily identified. It lays dormant, hiding in plain sight, in each and every architectural fragment that engages our senses and elevates the banality of daily life, for all of those who know where to look.

References:
1. For more on the NAi's forced merger, see Sergio Miguel Figueiredo, *Who Asked for a New Institute? A Farewell to the NAi*, August 20, 2013, http://www.domusweb.it/en/op-ed/2013/01/17/who-asked-for-a-new-institute-a-farewell-to-the-nai.html.
2. Piet Vollaard, *Meer Met Minder*, ArchiNed, June 10, 2011, http://www.archined.nl/nieuws/2011/juni/meer-met-minder/
3. NIADEC, *Creativiteit Als Noodzaak Beleidsplan 2013-2016 van Het Nederlands Instituut Voor Architectuur*, Design En e-Cultuur, February 2012, 8–9.
4. Samir Bantal et al., "The Storm Before the Calm" in *Architecture in the Netherlands: Yearbook 2009-2010* (Rotterdam: NAi Publishers, 2010), p. 7.
5. OASE, OASE *#67: After the Party: Dutch Architecture 2005*, Rotterdam: NAi Publishers, 2005.
6. Geert Bekaert and Christophe Van Gerrewey, "Architecture Devoid of Shadow" in *Rooted in the real : writings on architecture* by Geert Bekaert, Gent: WZW editions and productions, 2011, p. 39.

Sergio M. Figueiredo is an architect, author, curator and historian. He is an Assistant Professor of Architecture History and Theory at TU Eindhoven (TU/e), where he founded the Curatorial Research Collective (CRC), a fledgling curatorial and research group. At TU/e, he is also the coordinator for the research seminars on architecture and urbanism, as well as the chair and head curator of CASA Vertigo, the exhibition arm of the Department of the Built Environment. Figueiredo's work focuses on architectural institutions and exhibitions, particularly how they shape (and are shaped by) architectural culture, which he continues to develop through numerous contributions to publications and conferences, including guest editorships for the journals *OASE* and *Architecture & Culture*. His first book, *The NAi Effect: Creating Architecture Culture*, was published in 2016.

塞尔吉奥·M·菲格雷多是一位建筑师、作家、策展人和历史学家。他是埃因霍温理工大学（TU/e）建筑历史与理论的助理教授，在校内创办了策展研究小组（CRC）这一策展和研究团体。他还是该校建筑和城市主义研讨会的协调员，以及建成环境系展览部门CASA Vertigo的主席和首席策展人。菲格雷多的工作重点是建筑机构和展览，特别是它们如何塑造建筑文化（以及如何被塑造）。他也持续推动着建筑文化的发展，在出版物和会议方面作出了诸多贡献，包括在期刊《OASE》和《Architecture & Culture》担任客座编辑。他的第一本书《NAi效应：创造建筑文化》已于2016年出版。

NL Architects
Forum Groningen
Groningen 2007–2019

NL建筑师事务所
格罗宁根广场
格罗宁根 2007–2019

Forum Groningen is a new multifunctional building in the center of Groningen, a cultural "department store" filled with books and images that offers exhibition spaces, movie halls, assembly rooms, restaurants. The Forum aspires to become a platform for interaction and debate – a "living room" for the city.

Forum Groningen is not a library, not a museum, not a cinema, but a new type of public space where the traditional borders between these institutes will dissolve. Information will be presented thematically in a way that transcends the different media. The building is designed as a single clear volume to express the desire for synergy and to strengthen the shared ambition of combining different facilities into one new compound. A series of careful cuts joins the building to its site and generates a multitude of different appearances. Forum Groningen features an exceptional central space, an innovative atrium that with its horizontal "tentacles" forms the pumping heart of the venue. The void works as a spatial interface that binds all functions – movie theatre, book collection, expo, auditorium – and as such hopes to catalyze the exchange of knowledge and ideas.

A series of stacked "squares" emerges that can be experienced as the continuation of the network of open spaces in the city of Groningen. The vertical squares are publicly accessible and provide entries to the ticketed activities. The specific layout offers continuously changing views of the surrounding city and culminates in the roof terrace – a viewing platform and outdoor theater. Forum Groningen has been engineered "to accommodate finding not searching". The design stimulates exploration. It hopes to catalyze the desire to wander, to "browse" endlessly through a staggering interior landscape.

Credits and Data
Project title: Forum Groningen
Client: Municipality of Groningen
Location: Groningen, the Netherlands
Design: 2006 (Competition, 1st prize)
Completion: 2019
Architect: NL Architects, with ABT Engineering
Design team: (NL Architects) Pieter Bannenberg, Kamiel Klaasse, Walter van Dijk, Thijs van Bijsterveldt, Florent Le Corre, Sören Grünert, Iwan Hameleers, Sybren Hoek, Kirsten Hüsig, Mathieu Landelle, Zhongnan Lao, Barbara Luns, Gert Jan Machiels, Sarah Möller, Gerbrand van Oostveen, Giulia Pastore, Guus Peters, Jose Ramon Vives, Laura Riaño Lopez, Arne van Wees, Zofia Wojdyga, Gen Yamamoto with Christian Asbo, Nicolo Bertino, Jonathan Cottereau, Marten Dashorst, Rebecca Eng, Antoine van Erp, Tan Gaofei, Sylvie Hagens, Britta Harnacke, Jana Heidacker, Sergio Hernandez Benta, Johannes Hübner, Yuseke Iwata, Cho Junghwa, Linda Kronmüller, Jakub Kupikowski, Katarina Labathova, Ana Lagoa Pereira Gomes, Qian Lan, Justine Lemesre, Amadeo Linke, Fabian Lutter, Rune Madsen, Phil Mallysh, José Maria Matteo Torres, Victoria Meniakina, Shuichiro Mitomo, Solène Muscato, Lea Olsson, Pauline Rabjeau, Thomas Scherzer, Michael Schoner, Martijn Stoffels, Jasper Schuttert, Bartek Tromczynski, Carmen Valtierra, Elisa Ventura, Benedict Völkel, Vittoria Volpi, Murk Wymenga, Qili Yang, Yena Young, Alessandro Zanini
Project team: ABT Engineering (structural, building engineer, cost management), DGMR (building physics), Peutz (acoustics), Huisman en van Muijen (technical installations), NL Architects, De Munnik-deJong-Steinhauser, &Prast&Hooft, Tank (interior design)
Project area: 17,000 m² (building area), 10,000 m² (parking area)

pp. 192–193: Aerial view of the Forum Groningen. Photo by VolkerWessels Vastgoed. Opposite: View toward the book collection and the surrounding cityscape, from the atrium.

第 192-193 页：格罗宁根广场鸟瞰。对页：从中庭看向藏书区域及周围城市景观。

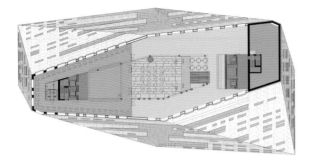

Tenth floor plan／十一层平面图

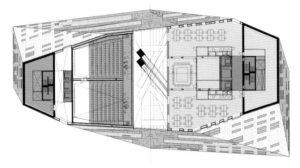

Seventh floor plan／八层平面图

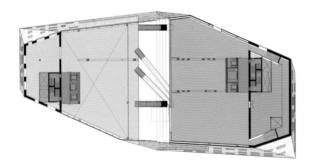

Third floor plan／四层平面图

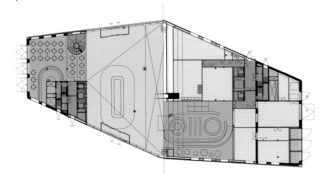

Ground floor plan (scale: 1/800)／一层平面图（比例：1/800）

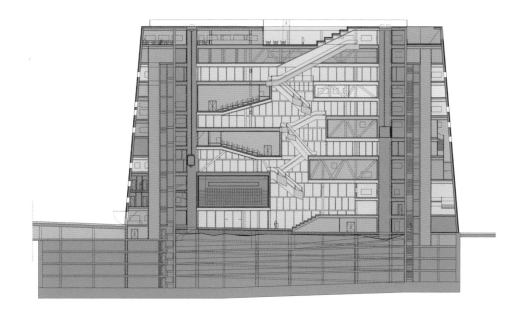

Section (scale: 1/800)／剖面图（比例：1/800）

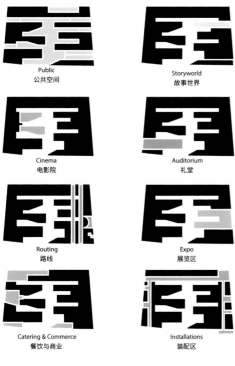

Public
公共空间

Storyworld
故事世界

Cinema
电影院

Auditorium
礼堂

Routing
路线

Expo
展览区

Catering & Commerce
餐饮与商业

Installations
装配区

Function diagrams／功能分析图

197

格罗宁根广场是格罗宁根市中心一座新型的多功能建筑。它是一个文化"百货店",里面装满了书籍和杂志,并提供了展览空间、影厅、会议室和餐厅。该建筑旨在成为一个城市客厅,搭建一个供人们交流和辩论的平台。

格罗宁根广场并不是一个图书馆或博物馆,也不是电影院,而是一种新型的公共空间。不同场所之间的传统界线被消除,不同主题将以一种跨媒介的方式在此展现。一个独立而清晰的建筑体,将不同设施整合在一个新的建筑中。该项目的设计充分体现了协同增效的理念。一系列经过谨慎考量的切割为建筑赋予了丰富的多面外观,并将建筑和场地紧密联系在一起。建筑内部有一个独特的中庭,新颖的空间和一系列水平延伸的"触角"仿佛是整个建筑跳动的心脏。中庭作为一个空间界面,将影院、藏书、展览馆和音乐厅联系在一起,并希望以此促进知识的交流和思想的碰撞。

由中庭连接的空间仿佛一系列堆叠的广场,这些广场成为格罗宁根城市开放空间网的延伸。这些纵向堆叠的公共广场也提供了收费活动空间的入口。建筑独特的布局提供了面向城市不断变化的视角,屋顶平台、景观台和室外剧场仿佛整个乐章的高潮。格罗宁根广场的设计旨在让使用者不断发现而非寻找。该设计将激发人们不断探索、不断漫游的欲望,让人们在令人惊叹的室内景观中持续游览。

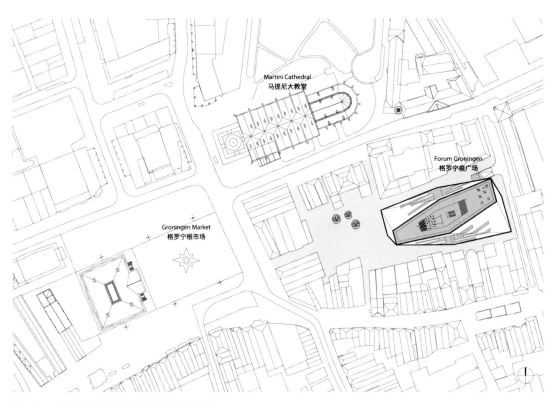

Site plan (scale: 1/2,500)/总平面图(比例:1/2,500)

p. 197, above: View from the atrium toward an exhibition area. p. 197, below: View showing the atrium's circulation and a social area under construction. Courtesy of De Zwarte Hond. Photos by Marcel van der Burg. Opposite: Aerial view of the forum in relation to the Martini Cathedral. Photo by Deon Prins.

第197页,上:从中庭看向展览区域;第197页,下:中庭交通空间和正在施工的社交区域。对页:广场和马提尼大教堂鸟瞰。

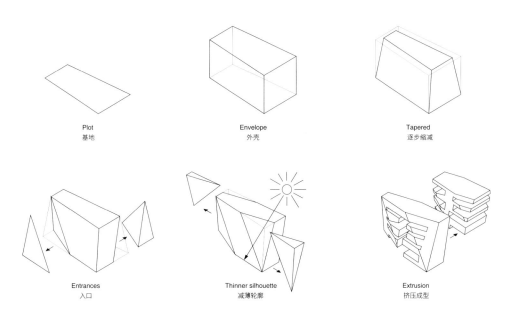

| Plot | Envelope | Tapered |
| 基地 | 外壳 | 逐步缩减 |

| Entrances | Thinner silhouette | Extrusion |
| 入口 | 减薄轮廓 | 挤压成型 |

Form studies／形体研究

Ector Hoogstad Architects
Utrecht Bicycle Parking
Utrecht 2019

埃克特·胡格斯塔德建筑师事务所
乌得勒支自行车车库
乌得勒支 2019

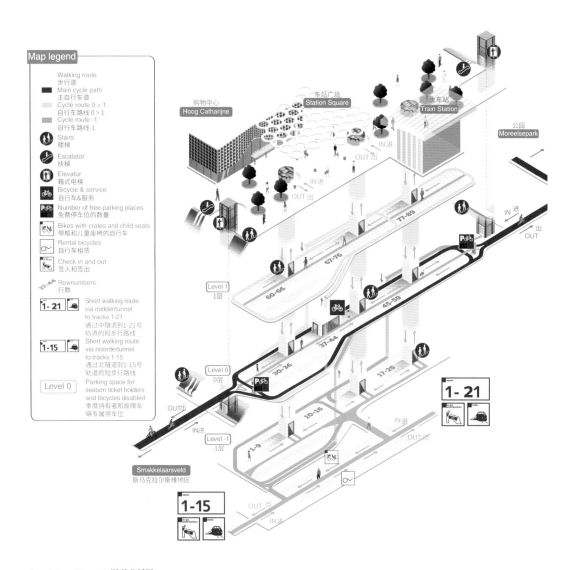

Circulation diagram／流线分析图

Historically, the Dutch have always been fervent cyclists. And as cycling is being discovered as a key ingredient of the sustainable city, this enthusiasm is growing even further. New bicycle typologies such as the introduction of the so-called e-bike, are helping to amplify this shift in. More and more public transport hubs will be complemented with extensive and user-friendly amenities for cyclists, as an increasing amount of people begin to favor the combination of cycling and public transport over cars.

Utrecht Central Station area is currently undergoing a major makeover. By healing the scars left behind by a number of "modernization" carried out in the 60s and 70s, and adding new functions to the area, this part of Utrecht is designated to finally become a vibrant and friendly part of the city center. To achieve this, inner-city highways are being replaced by more modest streets, and historic canals are being restored. The huge modernist megastructure that once glued the railway station and the Hoog Catharijne shopping mall – both the biggest of their kind in the Netherlands – was taken apart, thus allowing for a new public street and square to be inserted, along with a bicycle parking garage. This new "Stationsallee", a street mainly for pedestrians, ascends to a height of 6 m through a 30 m wide staircase and widens out into a square where the entrances of both the station and shopping mall are situated. An enormous iconic canopy marks the square and facilitates a sheltered crossing between the parts that were once connected.

The 3-story bicycle parking for over 12,500 bicycles is situated underneath the square. It was designed with 3 aims in mind: convenience, speed, and safety. In order to achieve this in a facility of this scale, cyclists are allowed to pedal all the way to their parking slots. The parking lanes branch off the cycle paths to ensure that users do not get in the way of cyclists passing through the system. Room for mounting and dismounting is located alongside the cycling lanes.

Modestly sloping ramps connect the parking areas on different levels. The walls are color-coded to indicate routing, and electronic signals indicate the position of free parking slots. Additional facilities such as a cycle repair shop, a cycle rental outlet and several floor managers meet its users' every need.

Stairwells and tunnels create direct connections to the elevated square, the main terminal building, and the platforms. Ensuring good orientation and plenty of daylight, the stairwells are located inside atriums covered by glass roofs. Large windows on the outer walls provide users with views toward the platforms and bus terminal.

The bike park uses durable materials such as concrete, steel, and chemically treated wood. Combined, these raw materials create an atmosphere in the building that still feels warm and pleasant. 3 concrete columns supporting the giant canopy extend all the way down into the parking area. Each of these trumpet-shaped elements is cast as a single element and has a diameter of 5 m at floor level, tapering to 1.2 m at the top.

The building is more than just infrastructure. It adds an exciting and surprising architectural dimension to the city. Cycling through the garage has become a unique experience – not just another part of everyday life in the city, but almost an attraction in its own right.

p. 201: View from the ground level. In the bicycle parking garage, traffic lines for pedestrians and bicycles are separated. Utrecht Station can be seen on the left. Opposite: Cyclist lane inside the bicycle parking area. Photos on pp. 200–207 by Petra Appelhof.

第 201 页：从一层看自行车库。自行车车库内，车道和步行道分开。左侧是乌得勒支车站。对页：自行车停车场内的骑行道。

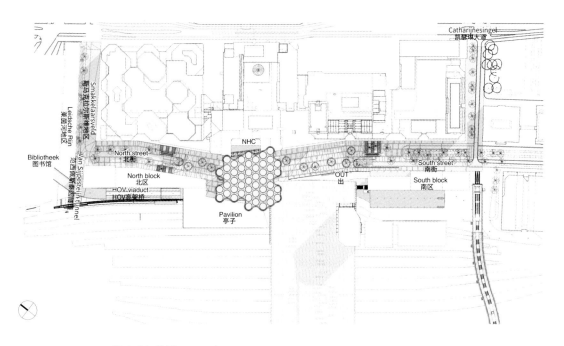

Site plan (scale: 1/3,500)／总平面图（比例：1/3,500）

历史上，荷兰人一直热衷于骑自行车。随着自行车被定义为可持续城市发展的重要组成元素之一，这股热情便更加高涨。新自行车种类的引入也在加速这种转型，所谓的电动自行车就是很好的例子。随着越来越多的人选择用自行车和公共交通来替代私家车，更多的公共交通枢纽将为骑行者提供友好、便捷的设施。

乌得勒支中央车站地区目前正在进行重大改造。除了解决 20 世纪六七十年代现代化建设所遗留的问题外，还计划增加新的功能，使该地区成为市中心一个充满活力和包容性的区域。为了实现这个目标，市区内的高速公路被更平缓的街道所取代，历史悠久的运河也正在被修复。曾经连接荷兰最大火车站和最大购物中心霍格·卡特琳娜的巨型现代建筑被拆除，为新的公共街道、广场以及自行车车库让出空间。这条新的"车站走廊"街道主要为行人服务，一座 30 米宽的楼梯将行人带到一个高于地面 6 米的广场上，车站和购物中心的入口都在这个广场上。广场上还有一个巨大的标志性顶棚，它代替了连接两栋建筑的结构，为行人提供了遮蔽。

自行车停车场则位于广场下方，共三层高，可容纳 12,500 辆自行车。该项目主要从方便、快捷和安全三个方面考虑进行设计。为了在这个规模下实现这样的目标，骑行者可以一直骑到自己的停车位。停车道与行车道分开，确保骑车人群和停车人群不会互相干扰。自行车道旁还留有安装和拆卸零件的空间。

各层停车区域以缓坡相连，墙壁上不同颜色的标识代表不同的路线，还有电子信号灯指示空车位的位置。除此之外，停车场还设有自行车修理店、自行车租赁店和楼层管理员，来满足使用者的各样需求。为了保证良好的建筑朝向和充足的自然光，楼梯间都位于有玻璃顶的中庭内，直接连接到上方的广场、车站和站台。使用者还可以通过外墙上的大窗户看到车站和站台。

这座自行车停车场使用了混凝土、钢材和经过化学处理的木材等耐久性材料。这些材料结合在一起营造出了一种温暖、令人愉悦的氛围。三根用来支撑巨大顶棚的混凝土立柱一直延伸到地下停车场。这些喇叭状的立柱底部直径为 5 米，顶部 1.2 米，每一根都是单独浇筑而成。

这座建筑不仅仅是城市基础设施，还为城市的建筑尺度增添了一份惊喜。它让骑行穿过停车场变成一种独特的体验。这座自行车停车场不仅仅是城市日常生活的一部分，它本身就是一个景点。

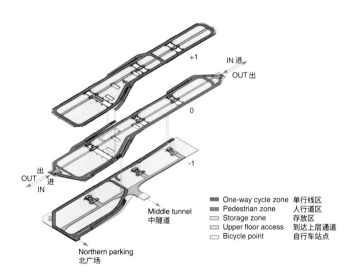

Bicycle route diagram／自行车路线图

Opposite, above: Parking garage. Users can stop by the reception desk while on a bicycle. Opposite, below: The columns that support the ground canopy continues to the underground parking garage.

对页，上：自行车停车场。使用者可以骑自行车经过前台；对页，下：支撑巨大顶棚的立柱一直延伸到地下停车场。

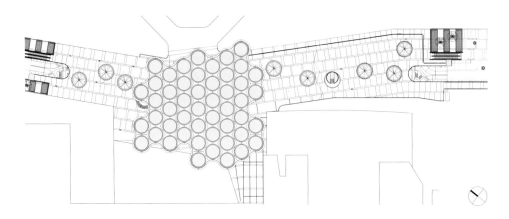

Roof plan／屋顶平面图

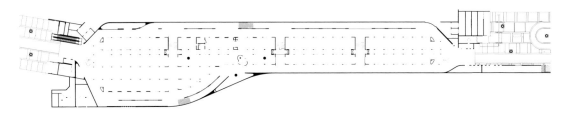

Basement bicycle parking floor plan (scale: 1/2,000)／地下自行车停车场平面图（比例：1/2,000）

Section (scale: 1/2,000)／剖面图（比例：1/2,000）

Credits and Data
Project title: Utrecht Bicycle Parking
Client: Gemeente Utrecht
Location: Stationsplein, Utrecht, the Netherlands
Completion: 2017 (first part of parking facility), 2019 (second part of parking facility)
Architect: Ector Hoogstad Architects
Design team: Joost Ector, Max Pape, Chris Arts, Stijn Rademakers, Gijs Sanders, Ralph Sijstermans, Lesley Bijholt, Romy Berntsen, Daniel Diez, Kees Bongers, Joost van der Linden
Project team: buro Sant & Co (landscape architect), Royal HaskoningDHV (construction), BAM Bouw en Techniek (builder)
Project area: 9,000 m² (site area), 23,473 m² (parking area)

This page: Ground level view of the canopy that covers the stairs serving as an entrance for pedestrians to access the parking garage.

本页：从一层看向覆盖自行车停车人行入口楼梯的顶棚。

Mecanoo
Delft City Hall and Train Station
Delft 2006–2017

麦肯诺建筑师事务所
代尔夫特市政厅和火车站
代尔夫特 2006–2017

Arriving in Delft is an unforgettable experience. From the outset, Mecanoo's idea was to design a station that makes it identifiable to visitors that they have arrived in Delft. The station, together with the new city hall, sits atop a new train tunnel built in place of the old concrete viaduct that divided the city into two since 1965. Coming up the escalators, the impressive ceiling with the historic map of Delft unfolds. And as you gaze outside, you will see the city and the old station as a contemporary version of Johannes Vermeer's painting "View of Delft".

Interweaving the past and future
The city of Delft reflects its past: the multitude of historic buildings and canals; the Prinsenstad city, closely connected to the Dutch Royal Family; and, of course, the world-famous Delftware ceramic factories. On the other hand, the Delft University of Technology is at the forefront of technical innovation. The character of Delft, epitomized in this combination of past and future, was the starting point for the design.

Delft Blue
A vaulted ceiling features an enormous historic 1877 map (see right) of Delft and its surroundings – connecting the station with the city hall. Within the station hall, walls and columns are adorned with a contemporary re-interpretation of Delft Blue tiles. You can walk directly from the station to the city hall. The glass skin of the building is designed to reflect the Dutch skies. Panels of fused glass with lens-like spheres reference a vernacular window design that can be seen throughout the historic city. The combination and rhythm of the open panels of high-performance glass and closed fused glass panels enable a high degree of energy efficiency.

Contextually compact
Throughout the design process, the building volume has been shaved and reformed to create a compact, highly efficient building form. The lowered roof lines at the corners provide a gradual transition towards the existing small-scale development of the Delft city center and the adjacent Wester Quarter. The building connects the historic inner city on the east side of the railway tunnel with the residential neighborhoods located on the west – realigning the center of Delft. Incisions in the glass volume of the city hall building form a pattern of alleyways and courtyards, which are themselves inspired by the intricate structure of Delft.

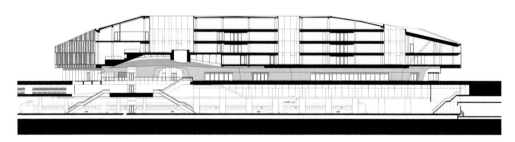

Long section (scale: 1/800)／纵剖面图（比例：1/800）

pp. 208–209: View of the station atrium. The Delft map printed onto a contemporary reinterpretation of the Delft blue tiles can be seen on the ceiling. Opposite, above: The entrance to the city hall can be seen on the left. Opposite, below: Interior view west of the station. Photos on pp. 208–219 courtesy of Mecanoo unless otherwise specified. p.212: The original map of Delft reflected on the station's ceiling. p.213: Reference of Delft blue.

第 208-209 页：车站中庭。天花板上的代尔夫特地图通过具有现代感的代尔夫特瓷砖得以展现。对页，上：左侧是市政厅入口；对页，下：车站室内西侧。第 212 页：车站天花板上是代尔夫特地图。第 213 页：代尔夫特蓝。

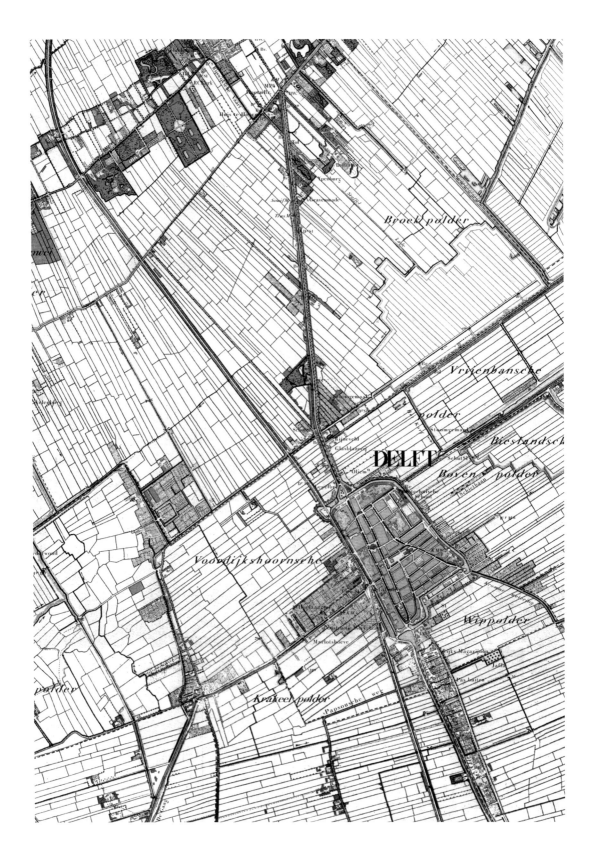

拜访代尔夫特是一个令人难忘的经历。一开始，麦肯诺的想法就是设计一个有足够辨识度的车站，让游客能立刻意识到自己已经抵达代尔夫特。车站和新市政厅坐落于一条新的火车隧道之上。这里曾是一座建于1965年的混凝土高架桥，自那时起这里就将城市一分为二。乘坐自动扶梯而上，天花板上的代尔夫特历史地图在视野中逐渐展开。向外望去，将会看到城市和旧的车站，仿佛画家约翰内斯·维米尔笔下的《代尔夫特风景》的当代再现。

过去与未来的交织

众多的历史建筑和静静流淌的运河，诉说着代尔夫特的历史。这座王子之城不仅和荷兰皇室家族有着紧密的联系，还有举世闻名的代尔夫特陶器工厂。除此之外，代尔夫特理工大学走在技术创新的前沿。代尔夫特就是这样一座过去与未来相互交织的城市，这也是整个设计的出发点。

代尔夫特之蓝

绘有巨大的1877年代尔夫特及周边地区地图的拱形天花板将车展与市政厅相连。车站大厅内，墙壁和立柱都用具有现代感的代尔夫特蓝色瓷砖来装饰。过往行人可以直接从车站走进市政厅。建筑的玻璃表皮反射着荷兰的天空。透镜状球体的熔融玻璃板，参考了当地传统建筑中常见的开窗设计。可开启的高性能玻璃面板和封闭的熔融玻璃板在立面上有韵律地交替，从而实现高效节能的效果。

紧凑的空间关系

在整个设计过程中，建筑的体量被适当削减和改造，以实现紧凑高效的建筑形态。转角处的屋顶稍有降低，和代尔夫特市中心及毗邻的韦斯特区现有的小尺度建筑形成自然的过渡。建筑将铁轨东侧历史悠久的内城和西侧的住宅区连接在一起，从而重新调整了代尔夫特的市中心。市政厅的玻璃体量上有几个切口，营造出了小巷和庭院的感觉，也呼应了代尔夫特错综复杂的城市结构。

Opposite: East facade. View from the canal on the east side. This page, above: Aerial view from the east. Photo by Harry Cock. This page, below: View of the bus terminal from south of the station and underground bicycle parking.

对页：从东侧运河边看建筑东立面。本页，上：东侧鸟瞰图；本页，下：车站南侧的公交车站和地下自行车停车区域。

This page, above: Courtyard on the second floor of the city hall. This page, below: View along the corridor on the second floor of the city hall. pp. 218-219: Office space.

本页，上：市政厅三层庭院；本页，下：市政厅三层走廊。第 218-219 页：办公空间。

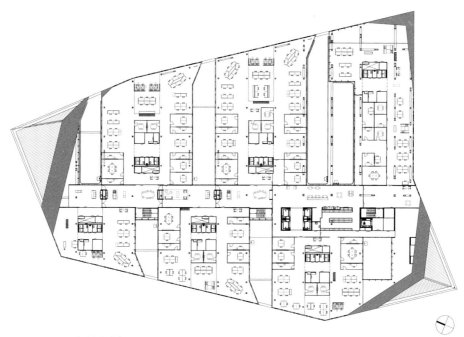

Third floor plan／四层平面图

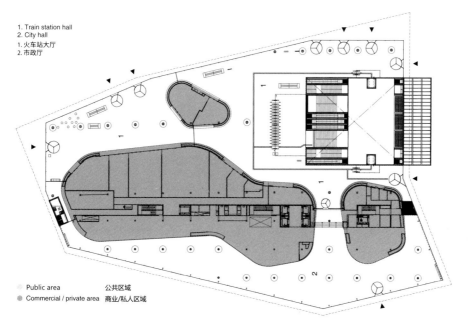

1. Train station hall
2. City hall
1. 火车站大厅
2. 市政厅

○ Public area　　公共区域
● Commercial / private area　商业/私人区域

Ground floor plan (scale: 1/900)／一层平面图（比例：1/900）

Credits and Data
Project title: Delft City Hall and Train Station
Client: Ontwikkelingsbedrijf Spoorzone Delft B.V.
Location: Markt 87, 2611 GS Delft, the Netherlands
Design: 2006–2010
Completion: 2012–2017
 (completion of station hall and the first phase of the city hall, 2015)
Architect: Mecanoo (main architect), Benthem Crouwel Architects
 (underground station platforms)
Project team: ABT B.V. (structural engineer), Deerns Raadgevende
 Ingenieurs B.V. (mechanical engineer), Basalt bouwadvies
 B.V. (cost consultant), LBP Sight (building physics, fire safety),
 Geerdes Ontwerpen (graphics ceiling)
Project area: 28,320 m² (total area), 19,430 m² (city hall)

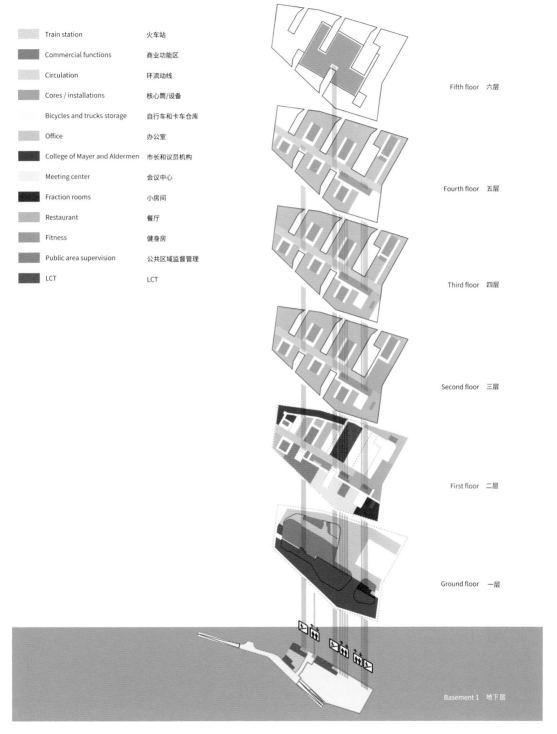

Exploded axonometric drawing／分解轴测图

Koen van Velzen Architects
Public Transport Terminal Breda
Breda 2004–2016

科恩·范·维尔森建筑师事务所
布雷达汽车站与火车站
布雷达　2004–2016

Credits and Data
Project title: Public Transport Terminal Breda
Client: ProRail with NS Stations and the city of Breda
Location: Stationsplein, Breda, the Netherlands
Design: 2004
Completion: 2016
Architect: Koen van Velsen Architects
Project team: Ballast Nedam with Hurks (contractor), Royal HaskoningDHV (structural engineer), Royal HaskoningDHV, Movares (installations engineer), LPB Sight, Nieuwegein (building physics, fire safety, acoustics)
Project area: 122,429 m²

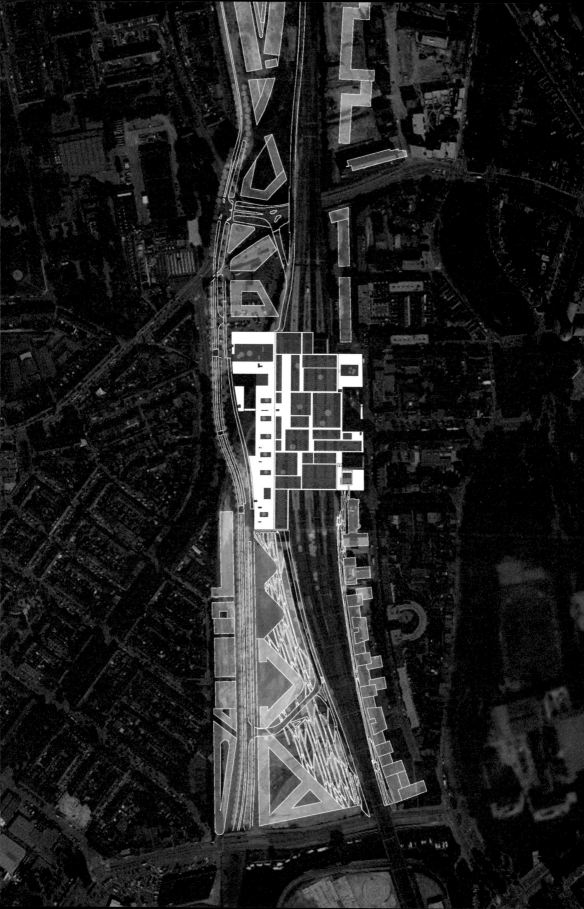

Section A／A剖面图

Section B／B剖面图

Section C (scale: 1/2,000)／C剖面图（比例：1/2,000）

The public transport terminal in Breda by Koen van Velsen architects is more than a train station. It is a new part of the city. An (inter)national train station and bus station, which also includes a wide diversity of other functions such as apartments, offices, retail, car parking, and bicycle parking, all combined into one single complex.

The public transport terminal is a spatial complex with a recognisable identity as a station – a meaningful and self-evident point of orientation in the city. The building is not designed as a multifunctional complex containing a public transport interchange, but as a station in which a substantial and complex program is combined into a single, strong recognisable form as an essential part of the city.

A building for the city

The complex is perfectly integrated as part of and for the city of Breda – characterful and firmly interwoven with the surrounding buildings. As a result, the public transport terminal is not a UFO that could have landed anywhere but is a specific solution designed to best suit the location and complex program.

The complex links 2 very different urban precincts on either side of the railway lines. Both on the north and south side, new public squares have been designed as an integral part of the project. The new public realm is an important factor in establishing a link to the existing city and forms the point of entry into the city of Breda.

The design puts the diversity of its users on the forefront – in particular, travellers. This results in clear and accessible spaces of appropriate allure, with a pleasant atmosphere, plenty of comfort, and maximum experience. The building is characterised by a multitude of perspectives, voids, and openings, which allow daylight deep into the building and enable the building to be experienced as a single coherent building volume. The multitude of perspectives and connections with the outside, create a building that is well-ordered and simple to navigate – making it an integral part of its surroundings.

A building for travellers

The train and bus platforms are combined into a single large central hall, which proves to be substantially more effective than a series of roofed platforms. Floors, walls, and ceilings are carefully adjusted to provide a coherent and spatial interior. The public semi-subterranean lobby connects directly to the train and bus platforms and links the two different city quarters on both sides of the railway lines. The roof of the station hall offers car parking facilities to travellers, residents, office workers, and is characterised by a large number of voids with trees. Within the complex, each function has its own individual character which is mostly reflected in the variation in materials and detailings used. This results in easy way-finding and identification of the different functions in the complex.

Urban development

The public transport terminal is the center point of an ambitious urban development in Breda where a former industrial area and marshalling yard is transformed into a vibrant new urban quarter. Officially opened on 8 September 2016, the transformed Breda Station – together with Rotterdam, Arnhem, Utrecht, The Hague, and Delft Station – forms as part of the Nieuwe Sleutel Projecten (NSP, New Key Projects).

p. 220: Aerial view from the east. p. 221: Plan view of the proposed public transport terminal. pp. 222–223: View from along the rooftop car park. p. 224-225: View from the residence buiwvlding looking towards the underground entrance that leads to the bicycle parking and station platform. Opposite: View from the first floor level looking towards the residential building at the back. Photos on pp. 220–231 by René de Wit.

第220页：东侧鸟瞰图。第221页：拟建公共交通枢纽的平面图。第222-223页：屋顶停车场。第224-225页：从住宅楼看向地下自行车停车入口和站台。对页：从二层看向背面的住宅楼。

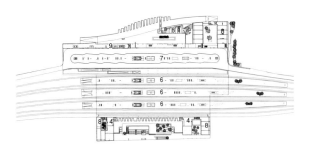

First floor plan／二层平面图

Sixth floor plan／七层平面图

1. Entrance
2. Public square
3. Station hall
4. Retail, restaurant
5. Transfer lobby
6. Train platform
7. Bus platform
8. Housing
9. Office
10. Bicycle parking
11. Patio
12. Car parking roof

1. 入口
2. 公共空间
3. 车站大厅
4. 零售、餐厅
5. 换乘大厅
6. 列车站台
7. 公交站台
8. 住房
9. 办公室
10. 自行车停车库
11. 院子
12. 屋顶停车场

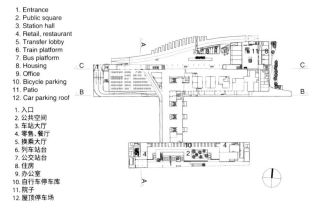

Ground floor plan (scale: 1/5,000)／一层平面图（比例：1/5,000）

Fifth floor plan／六层平面图

Opposite: View overlooking the patio within the residential building, northwest on site.

对页：俯瞰西北角住宅楼中庭。

这座由科恩·凡·威尔森建筑师事务所设计的布雷达汽车站与火车站，是一个超越了车站功能的城市新元素。它不仅是国内外火车和汽车的交通枢纽，更是一个囊括了集合住宅、办公楼、商铺、屋顶停车场、室内外自行车停放处等多种空间功能的综合设施。

这个交通枢纽是一个具有明确车站身份的空间综合体。所以，在大家心目中它是不言而喻的城市起点，也很容易理解。在设计构思上，这座建筑以一个不可或缺的强有力的单体形态，来组织起错综复杂的各功能项目，而并非一个统合公交节点的多功能综合体。

为城市而建的建筑

这座综合体是为布雷达城市而存在的，与城市浑然一体。既葆有自己的特点，又与临近的建筑无违和地相交织在一起。也因此，这个枢纽并不是随处出现的UFO（不明物）形象，而是与该场所复杂的功能需求相吻合的设计。

建筑位于轨道线路两侧，连接起了全然不同的两个区，并作为项目的一部分，在站南站北各设计了一座城市广场。新公共空间既是建立与原城市相联系的重要元素，也是通向布雷达的城市入口。

新车站的设计关注以游客为主的不同使用人群。建筑空间明快，易达，富有魅力；建筑氛围舒畅愉悦，兼具充盈感和舒适性，这些都为使用者提供了绝佳的体验。

阳光深入内部，人们可以体验到大型设施一体化，建筑还拥有多种视角（透视角度）、通风口和开口，这些赋予建筑鲜明的特征。多视角与外部相接，使建筑整体秩序井然，通达便捷。这也使建筑融入周围环境成为可能。

为游客而造的建筑

火车和巴士站台在一个巨大的中央大厅相连接。不言而喻的是，这比带屋顶的站台并列的形式更具优越性。地板、墙壁和天花板都经过精心调整，使整个室内空间更加和谐连贯。半地下的公共大堂直接连接到火车和巴士站台，并将铁轨两侧两种不同的城市空间联系起来。车站大厅的屋顶是游客、住户和办公人员的停车区域。大量树木从建筑留空处伸向屋顶，给这个空间赋予了特色。每个功能空间的特色都通过材料的选择和细节的设计以区分，这也使得人们能够更容易地在建筑中识别不同的功能区域。

城市发展

布雷达汽车站与火车站是整个城市发展的核心，将以前的工业区和调度站转变为一个充满活力的城市新区。自2016年9月8日正式开放以来，布雷达汽车站与火车站与鹿特丹、阿纳姆、乌得勒支、海牙和代尔夫特车站一起组成了荷兰重点新建项目的一部分。

Opposite, clockwise from top left: North facade view from the observatory above the bicycle parking. Station platform. Bus terminal. The commerce and restaurant area on the first floor. Transfer lobby on the ground floor. Stairs leading from the ground floor to the bus terminal on the first floor. Public square on the ground floor. Passage way along the bicycle parking on the ground floor.

对页，从左上顺时针方向依次为：从自行车停车上方瞭望台看北立面；车站站台；公交巴士站；二层商业餐饮区；一层换乘大厅；楼梯从一层通往二层巴士站；一层公共广场；一层沿自行车停车场的人行通道。

Cruz y Ortiz Architects
The Rijksmuseum
Amsterdam 2001–2013

克鲁兹·奥尔蒂斯建筑师事务所
荷兰国家博物馆
阿姆斯特丹 2001–2013

The Rijksmuseum in Amsterdam was designed in the late nineteenth century by Dutch architect Pieter Cuypers. The function of the building was twofold: one was the national museum, and the other was the gateway to the south of Amsterdam.

The museum has paid an overly high price for its urban role as a connecting element between what was then the existing city to the North, and the newer developments towards the South. A walkway – virtually a street – runs through the building from North to South splitting it into 2 parts, necessitating 2 entrances – both towards the North – and 2 main staircases. This means that only on the first floor, the Eastern and Western parts of the building are joined; on the ground floor and basement, they are divided. The need for an exhibition space meant building within the courtyards, which led to a lack of natural light. This also brought about a kind of labyrinth where the visitor is given no information concerning their whereabouts.

The intervention on the building was, firstly, meant to open up a new and unique entrance to the museum admission in the central passage hall. Secondly, it was to recover the courtyards and exhibition spaces, regaining somewhat their original state, or at least their dimensions.

The large space, generated by opening and connecting the courtyards, houses all essential uses for visitors and offers a suitable space on a scale that the grandeur of the building deserves. One enters this hall from the passageway, and tours to the exhibition areas start at this point, linking with the original grand stairs. Natural limestone is used in this new space created. Connected under the passage in each of these courtyards is a structure – with an acoustic and lighting purpose – called the "chandelier" suspended from above.

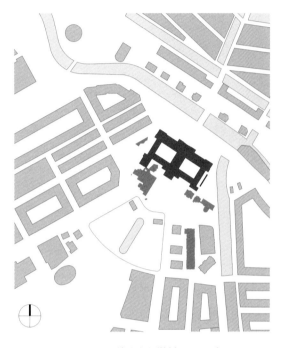

*Site plan (scale: 1/8,000)／*总平面图（比例：1/8,000）

pp. 232–233: View of the public walkway through the museum. The Rijksmuseum serves as a connection between the southern and northern part of Amsterdam. Photos on pp. 232–233, 236–245 by Duccio Malagamba. Opposite, above: Photo of the north elevation taken in 1960. Opposite, middle: Photo of the public walkway taken in 1936. Opposite, below: Interior view of the western courtyard in 1910. Photos on opposite page courtesy of the Rijksmuseum.

第 232-233 页：博物馆入口通廊。荷兰国家博物馆是阿姆斯特丹南北两侧的连接枢纽。对页，上：北立面，照片拍摄于 1960 年；对页，中：公共走廊空间，照片拍摄于 1936 年；对页，下：西侧中庭内部，照片拍摄于 1910 年。

Credits and Data
Project title: The Rijksmuseum
Client: Programmadirectie Het Nieuwe
Location: Rijksmuseum Museumstraat, Amsterdam, the Netherlands
Design: 2001
Completion: 2013
Architect: Cruz y Ortiz Architects
Design team: M. Huisman, T. Offermans, T. Reventós, O. García de la Cámara, M. Ter Steege, A. López, J. L. Mayén, C. Hernández, A. Vila, V. Bernícola, J. Kolle, V. Breña, S. Gutiérrez, M. Pelegrín, I. Men- nenga, J. Pérez-Goicoechea, L. Gutierrez, C. Arévalo, R. Peinado, J.C. Mulero, M. Velasco
Project team: HMADP Architects (local architect), Arup (lighting design), Copijn Tuin - en Landschapsarchitecten (landscape architect), Van Hoogevest Architects (restoration architect), Cruz y Ortiz Architects, Indigo (digital imaging), Jacinto Gómez (model), ARCADIS (structural engineer), Arup, Royal HaskoningDHV, Van Heugten (technical engineer), DGMR (fire advisor), Cruz y Ortiz Architects (construction management), JP van Eesteren, BAM, Homij, Kuipers, Unica, Woudenberg Koninklijke, Moehringen (contractors), Van Hoogevest, Cruz y Ortiz Architects, Rijksvastgoedbedrijf (work control)
Project area: 16,000 m² (site area), 45,000 m² (building area)

荷兰国家博物馆由荷兰建筑师皮耶特·库佩斯于19世纪下半叶设计而成,这座建筑身兼两职——既是国家博物馆,同时也是通往阿姆斯特丹南部的门户。

这座博物馆是已发展成熟的荷兰北部和新开发的荷兰南部的交接点,这使她为其扮演的历史角色付出了很大的代价。一条人行通道贯穿建筑南北,博物馆被一分为二,于是,博物馆需要两个北向的入口和两座主楼梯。这也意味着建筑东西两部分仅在二层相连,在一层和地下层都是分开的。此外,因展览空间所需,在中庭也建有建筑,这导致自然光无法照射进来。来馆者因此常不知自己身在何处,这里俨然成为了一个"迷宫"空间。

基于此,博物馆的改建,第一目标是在中央通廊上设置一个新的博物馆专用入口;第二目标是将中庭和展览空间改回来,尽可能恢复到原来状态,至少也要恢复到原来的规模。

通过打开中庭从而产生的这个连起来的大空间,不仅提供了来馆者参观所需的所有功能,并为这座壮观的美术馆带来了与其极为相契的恢宏的空间尺度。人从通廊进入大厅,从与原有大台阶相连的地方开始,就可以到达展览区,开启参观之旅。新建的空间里,使用了天然石灰石。中庭在通廊下方相连,其上方垂吊着一组能同时固定音响和照明设备的、被称为"枝吊灯"的结构体。

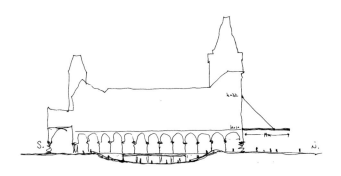

pp. 236-237: South elevation. This page: Concept sketches by the architect. Opposite: Entrance into the gallery area. pp. 240–241: Interior view of the west courtyard. p. 242: Interior view of the gallery.

第 236-237 页:南立面。本页:概念手绘图。对页:进入画廊区域的入口。第 240-241 页:西侧中庭的内景。第 242 页:画廊的内景。

1. Museum passageway	6. Shop	1. 博物馆通廊	6. 商店
2. Control point	7. Cafe	2. 控制中心	7. 咖啡厅
3. Auditorium lobby	8. Chandelier	3. 礼堂门厅	8. 吊灯
4. Auditorium	9. Velum	4. 礼堂	9. 剧场帐篷
5. Meeting room	10. Exhibition space	5. 会议室	10. 展览空间

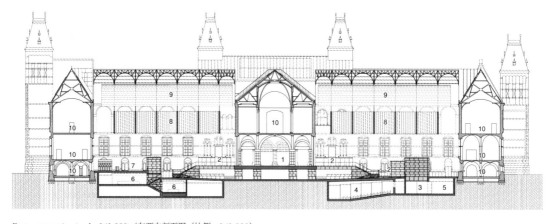

East-west section (scale: 1/1,000) ／东西向剖面图（比例：1/1,000）

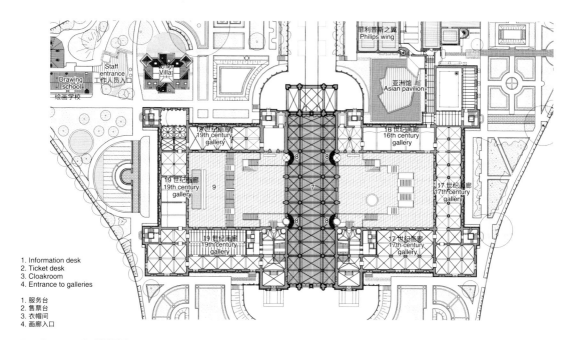

1. Information desk
2. Ticket desk
3. Cloakroom
4. Entrance to galleries

1. 服务台
2. 售票台
3. 衣帽间
4. 画廊入口

First floor plan／二层平面图

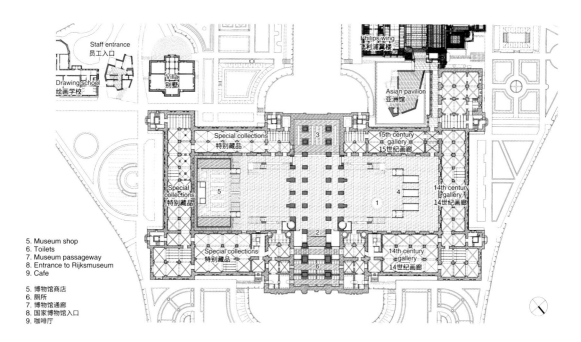

5. Museum shop
6. Toilets
7. Museum passageway
8. Entrance to Rijksmuseum
9. Cafe

5. 博物馆商店
6. 厕所
7. 博物馆通廊
8. 国家博物馆入口
9. 咖啡厅

Ground floor plan (scale: 1/1,500)／一层平面图（比例：1/1,500）

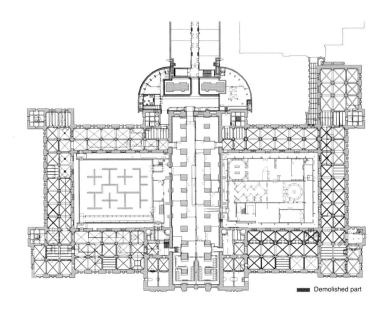

■ Demolished part

Original ground floor plan／改造前的一层平面图

1. Connection to truss (existing structure)
2. Deformable U-profile
   (different settlement between old and new structure)
3. Round steel bar
4. Steel profile 100×245×10 mm
5. Steel profile L=shape 150.100.10
6. Lighting cable
7. Wooden piece DM t=18 mm painted in RAL 9010
8. Wooden acoustic panel
   84×336 mm painted in RAL 9010
9. Folded aluminium sheet painted in RAL 9010
10. Laminated wood (reinforcing the aluminium sheet)
11. Spot lighting
12. Wooden structure of 40×60 mm
13. Mineral wool insulation t=60 mm
14. LED-lighting
15. Aluminium profile

1. 桁架连接（现有结构）
2. 可变形的 U 型剖面（新旧结构间的不同沉降）
3. 圆形钢筋
4. 钢型材料 100×245×10 mm
5. 钢型材料 L= 形状 150.100.10
6. 照明电缆
7. 木件 DM t=18 mm 涂在 RAL 9010 中
8. 84×336 毫米木质隔音板，RAL 9010 涂漆
9. RAL 9010 涂漆的折叠铝板
10. 层压木材（加固铝板）
11. 现场照明
12. 40×60 mm 的木结构
13. 矿棉绝缘 t=60 毫米
14. 发光二极管照明
15. 铝型材

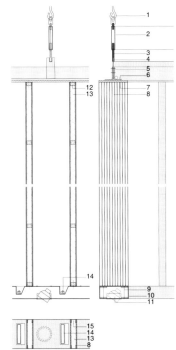

Detail of "the chandelier" (scale: 1/45)
"枝吊灯"细部详图（比例：1/45）

1. Natural stone paving Gascogne Blue t = 20 mm on mortar grip
2. Concrete compression layer t = 100 mm with floor heating
3. Extruded polystyrene insulation t = 40 mm
4. Reinforced concrete structure t = 250 mm
5. Concrete compression layer t = 50 mm
6. Extruded polystyrene insulation t = 100 mm
7. Steel containment screens 11 m long
8. Metal profile on pavement joint
9. Structural joint
10. Belgian natural stone paving on mortar grip
11. Foundations of existing structure
12. Oak flooring t = 6.3 mm on wooden board t = 8 mm
13. Calcium sulphate panels of 600×600×4mm on plots
14. Wooden porch h = 2,400 mm
15. Plasterboard panel t = 15 mm on subframe
16. Natural stone finishing Gascogne Blue t = 30 mm anchored to concrete structure
17. Aluminium folded sheet t = 3 mm for general ligthing
18. MDF wooden lintel t = 20 mm on subframe
19. Reinforced concrete beam 400×700 mm
20. Railing steel bars t = 8 mm and glass 6 + 6 mm
21. False ceiling plasterboard panel t = 15 mm
22. Concrete slab t = 180 mm
23. Reinforced concrete beam 250×900 mm into existing structure
24. Preexisting structure
25. Expansion joint
26. Anti-slip bushhammered
27. Natural stone Gascogne Blue t = 150 mm
28. Low walls of brick
29. Belgian natural stone paving on mortar grip
30. Existing terrazo flooring
31. Sandwich panel (fire resistance 30 min.:Interior insulation of rockwool t = 72 mm fibercement panel t = 15 mm plasterboard panel t = 7 mm

1. 天然加斯科尼蓝石材铺装 t=20mm，砂浆基层
2. 混凝土压缩层 t=100mm 带地暖
3. 挤压成型的塑聚苯乙烯隔热材料 t=40mm
4. 钢筋混凝土结构 t=250mm
5. 混凝土压缩层 t=100mm
6. 挤压成型的塑聚苯乙烯隔热材料 t=100mm
7. 钢密闭风长 11m
8. 路面金属填缝
9. 结构接头
10. 天然比利时石材铺装，砂浆基层
11. 现有结构的基础
12. 橡木地板 t = 6.3mm，下层为木板 t = 8mm
13. 基地上覆 600x600x4mm 硫酸钙板
14. 木门廊 h = 2400mm
15. 辅助构架上的石膏板面板 t=15mm
16. 天然加斯科尼蓝石材饰面层 t= 30mm，固定在混凝土结构上
17. 用于通常照明的铝折页 t = 3mm
18. 辅助构架上的中密度纤维木板过梁 t = 20mm
19. 钢筋混凝土梁 400x700mm
20. 钢栏杆 t = 8mm，玻璃 6 + 6mm
21. 石膏板吊顶天花板 t = 15mm
22. 混凝土板 t = 180mm
23. 250x900mm 钢筋混凝土梁与现有结构相连
24. 现有结构
25. 伸缩缝
26. 防滑荔枝纹
27. 天然加斯科尼蓝石材 t=150mm
28. 矮墙砖
29. 天然比利时石材铺装，砂浆基层
30. 夹水磨石地板
31. 夹芯板（耐火 30 分钟：岩棉内部保温层 t=72mm 纤维水泥 t=15mm 石膏板 t=7mm）

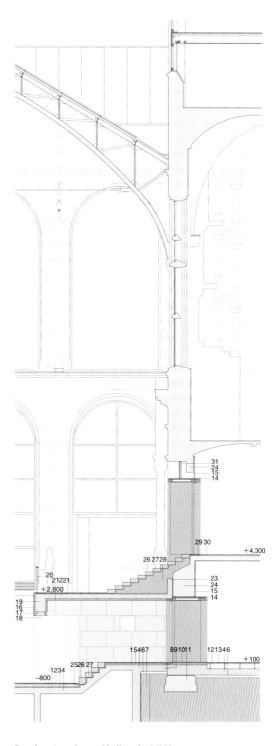

Detail section of central hall (scale: 1/150)
中央大厅剖面详图（比例：1/150）

An Interview with Floris Alkemade:
The Future of the Netherlands
Kirsten Hannema

采访弗洛里斯·阿尔克马德：

荷兰的未来

克丝汀·汉内马

Summary of the conversation between Kirsten Hannema and the Chief Government Architect Floris Alkemade sharing his thoughts on the need for change, imagination and narratives.

The crisis. "If one subject has been decisive for the past 10 years of architecture in the Netherlands, then that's it," says Floris Alkemade. The most severe crisis since World War II, which spread from the United States to Europe in 2008 and drew a line under twenty years of unprecedented prosperity in Dutch architecture.

"When I started at OMA in 1990, it seemed as if the only way was up, with market parties positioning themselves in a way that we could do all sorts of large projects, such as Euralille, which I was in charge of," Alkemade continues, "It resulted in a generation of architects 'that only went well', but turned out to be vulnerable, because 'their relevance was focused on extravagance'."

Never waste a good crisis
Alkemade believes one should never waste a good crisis. He looks back at an interesting and fruitful period. It struck him that "incredibly good architecture has been realized". As an example, he refers to the LocHal (See pp. 122–133) in Tilburg – a former locomotive shed of the Dutch National Railways – transformed by CIVIC Architects, Braaksma&Roos, Inside Outside and Arup into a public library that doubles as City Campus.

Alkemade continues to describe the project. "Before the crisis the municipality intended to demolish the buildings on this site to build new housing, but as the financing for that plan was no longer possible, it was canceled. People started to think about other possible functions, all of a sudden they looked much more careful: what can you do with such a building? Not from an economical driver, but from a moment of vulnerability, a new way of thinking emerged, about sustainable reuse, co-creation, creating public space. With a fantastic end result."

The LocHal tells in a nutshell what being Chief Government Architect is all about for Alkemade, convincing the Netherlands – vulnerable because of its location below sea level – of the necessity to follow a new, sustainable road, driven by the imaginative and narrative force of design.

Architecture as luxury
By now the economy is back on track, but this does not automatically account for architecture offices. From Alkemade's perspective, "Architects are building again, but differently. The use of BIM (Building Information Modeling) software for example, has taken a huge flight under constructors, who use it to draw out architectural designs themselves. After the preliminary design had been completed, they tend to see the work of architects as a luxury, and not a necessary condition to realize a project. The architect is pushed back in the role of aesthetic consultant, assignments are more and more reduced to obtaining an environmental permit. Meanwhile, under the influence of neo-liberal policies, the government is increasingly allowing market parties to build, and less actively raises the public interest of architecture. Since the 1990s architects have been assimilated with the market in such a way

（首席政府建筑师弗洛里斯·阿尔克马德在与克丝汀·汉内马的对谈中，分享了他对改变、想象和叙事的必要性的看法。以下是那次谈话的摘要。）

金融危机。"如果说过去十年中，有什么对荷兰建筑产生了决定性的影响，那就是它"，弗洛里斯·阿尔克马德这样说道。这场自"二战"以来最严重的危机，在2008年从美国蔓延至欧洲，为荷兰建筑持续20年的空前繁荣画上了一个句号。

"1990年我加入OMA时，似乎前景一片大好，从市场情形来看，我们可以接下所有类型的大项目，包括我负责的欧拉里尔购物中心"，阿尔克马德继续说："但事实上，'一帆风顺'的一代建筑师并非坚不可摧，因为'他们的主要任务是表现奢侈'。"

机不可失

阿尔克马德相信，良机不可失。他回忆起曾经硕果累累的岁月，感慨那时"绝妙的建筑都能够实现"。他举例了位于蒂尔堡的劳克哈尔图书馆（LocHal）（见122-133页）。它的前身是荷兰国家铁路的机车库，在城市建筑事务所、布拉克斯马和鲁斯建筑师事务所、内外建筑师事务所、奥雅纳的共同改造下，成了一座公共图书馆兼"城市校园"。

阿尔克马德说起项目的来龙去脉："金融危机开始前，相关部门计划拆除这些建筑，再建造一些新的，但由于资金无法到位，计划搁浅了。于是人们开始挖掘这一地块的更多潜力，瞬间开始观察入微，思考能够为老建筑做点什么。正是在这番困境之中，而非经济景气的时刻，人们关注起可持续、共创、公共空间的营造。最终的结果实在妙不可言。"

在阿尔克马德看来，劳克哈尔图书馆精准反映了首席政府建筑师的角色定位，也让荷兰这个低于海平面的国家相信，有必要借助想象力和叙事力，走上一条全新的、可持续的道路。

建筑作为奢侈品

如今经济已经回到正轨，但建筑事务所的情况尚未得到保障。阿尔克马德认为，"建筑师再次开始建造，但方式有所不同。举个例子，BIM（建筑信息模型）软件在施工方中迅速普及。他们使用BIM自己绘制方案、完成初步设计，因此往往觉得建筑师的工作是奢侈的锦上添花，而非实现项目的必要条件。于是建筑师被推回到美学顾问的位置，更多的委托简化到了只需要为项目拿到环境许可。与此同时，在新自由主义政策的影响下，政府逐渐允许市场各方自行建造，对于增进建筑的公共利益愈发懈怠。1990年代以来，建筑师都在市场的浪潮中随波逐流，不再书写自己的故事。"

因此，阿尔克马德将恢复建筑的质量和意义视为己任。"塑造美固然重要，但我们要做的远不止于此。"他看到了政治家和市场都不能解决的社会问题：（经济型）住房的严重短缺，日益增长的医疗需求和成本，人口流失的乡村，必须实行可持续发展的农业。"这些都是关乎空间的问题，需要长远的眼光和想象。也正是这种工作方式，能够给专业团队和建筑思维带来新的价值定位"，他评论道。

公开征集提案

首席政府建筑师属于政府的独立顾问，没有任何权力。为了引起公众对上述问题的关注，阿尔克马德决定公开征集解决方案。2015年欧洲难民危机最高峰时，他发起了以"家外之家"（A Home away from Home）为主题的设计竞赛，为持有居留许可者、学生、外籍人士等解决住房问题。胜出

that their own story has been lost."

So there's his mission: to bring back quality and relevance of architecture. "Aesthetics is important, but our profession is about much more than that." He sees social issues to which politicians and the market do not have adequate answers: the acute shortage of (affordable) homes, the growing demand – and costs – for health care, the countryside where the population is moving away and agriculture must be made sustainable. "These are spatial tasks that require a long-term vision and imagination. At the same time, it is a way of working that gives the professional group and architectural thinking a new position," he comments.

Open Calls

The role of the Chief Government Architect is that of independent advisor to the government; he has no mandate. To draw attention to the aforementioned issues, Alkemade decided to organize open calls. In 2015, at the height of the refugee crisis in Europe, he launched the design competition *A Home away from Home*, focused on housing solutions for status holders and other home seekers such as students and expats. The winning designs varied from a styrofoam "igloo" to a 3D-printed building system for interiors with which vacant office buildings temporary can be used as homes.

In 2016, followed the open competition *Who Cares*, around the question of how we can better organize neighborhoods for the growing group elderly people – starting from the idea that a city that is well equipped for the elderly and the vulnerable is a good city for everyone. *With Brood en Spelen (Bread and Circuses, 2018)*, he looked for "promising and innovative ideas for rural renewal". Like the plan by Studio Marco Vermeulen to plant fast growing poplar woods for the production of cheap laminated building wood, as an alternative for (environmentally harmful) concrete and steel.

As it turns out, it is difficult to get the winning plans off the ground. "We want to put a different way of thinking on the agenda, it's about the experiment", explains Alkemade, "Innovation is hidden in the barriers that we create; making concessions means you lose them. However, we have refined the system so as to increase the chance of realization. For *A Home away from Home*, we only invited designers. With *Who Cares*, we asked parties from the health care sector to join design teams. For Brood en Spelen, we went one step further by linking landowners to design teams who would bring in a piece of land to build on."

Panorama Netherlands

"The chance of success increases if you have an overall vision," says Alkemade. After all, themes such as housing, care, the city, the countryside, energy transition and climate adaptation cannot be viewed separately. Together with his fellow Government Advisors for the physical environment, landscape architect Berno Strootman and architect/urban designer Daan Zandbelt, at the end of 2018, Alkemade launched *Panorama Netherlands* – a future perspective for the spatial planning of the Netherlands. It comprises a publication, application, and traveling exhibition that "shows how major issues of today can be the key to

的方案各具特色,既有泡沫塑料制成的"冰屋",也有使用3D打印技术的室内建造系统,后者便于将空置办公楼改为临时住房。

2016年,阿尔克马德继续举办了题为"谁来关爱"(Who Cares)的公开竞赛,试图为日益壮大的老年群体创造更好的社区环境。该提案是基于这样的理念:一座老弱群体宜居的城市,才是人人共享的好城市。"面包和游戏"(Brood en Spelen, 2018)的竞赛则是为了征集"有望实现乡村更新的新思路",其中马可·维梅伦工作室提出种植快速生长的杨树,制造廉价的胶合木,替代有损环境的钢筋混凝土。

然而事实证明,胜出的方案很难落地。"我们想换个角度看待这一问题,把它当作一场实验",阿尔克马德解释说:"创新会被我们自己创造的屏障所阻碍,作出让步,就意味着与创新失之交臂。但我们还是改良了竞赛的架构,希望增加方案实现的可能性。在'家外之家'中,我们只邀请了设计师;到了举办'谁来关爱'时,我们开始让医疗人士加入设计团队;'面包和游戏'则更进一步,为土地所有人和设计团队搭起了一道桥梁。"

全景荷兰

"如果具备了宏观的视野,距离成功就更近了一步",阿尔克马德说道。归根结底,住房、医疗、城乡、能源转型、气候适应等议题都不是孤立的存在。2018年年底,阿尔克马德与他的两位同事,景观建筑师伯尔诺·斯特罗曼和建筑/城市设计师达恩·赞贝尔特(两位也是政府的物理环境顾问),启动了"全景荷兰"项目,展望荷兰未来的空间规划。这个项目以出版物、应用程序、巡回展览的形式呈现,"介绍今天的重大问题如何决定性地影响未来的结构性变革"。从8米长、360°环绕的展出内容来看,2050年的荷兰将在北海(大西洋)建有风电场,海岸线得到加固,城市的绿化令人惊叹。

在荷兰最大的日报《人民报》上,专栏作家伯特·瓦根多普用一个词概括了"全景":乐观主义。在他看来,这与"半个荷兰沉在水中、另外一半内战肆虐的世界末日前景"形成了鲜明对比。瓦根多普将这种乐观主义称为"一缕清风",并希望政治家能够"听取他们的建议"。但与此同时,他仍然半信半疑,因为"无论左翼还是右翼,抬头的民粹主义至少有一点共通之处,即不可救药的悲观主义"。

刚健质朴

这样的负面情绪也妨碍着阿尔克马德的另一个重大项目:海牙国会议事堂(荷兰下议院所在地)的翻新。整个建筑群历史悠久,耐火性差,极其缺乏维护,因而需要全面更新设施。阿尔克马德将此次"技术性"翻新(预计耗资4.75亿欧元)视为改善建筑空间的良机。在他的提议下,12位建筑师被指定负责各个区域的翻新,建筑师莉斯贝思·凡德波(DOK)与艾伦·凡卢恩(OMA)担任总体协调和监督者。

然而,本应属于国家机密的设计过程始终存在问题。在初步设计方案被泄露给媒体之后,议员们纷纷作出"浮夸狂妄"的评价,导致DOK和OMA两家建筑事务所被迫退出,令阿尔克马德颇为沮丧。"决定事态发展的,不是内容,而是形象。如果艺术家创造的空间看上去与环境格格不入,一些人就认为需要耗费过多的税金。每个显而易见的改变,都会被视为负面因素",他说。

"形象本身就带有负面色彩",阿尔克马德进一步解释道:"事实证明,议员要读懂图纸并不容易,因此很难估量方案的价值。但这也与文化有关,在我工作过的法国和

welcome structural improvements in the future." An 8-meters long and 360-degree perspective offers an impression of the Netherlands in 2050, with windmill parks in the North sea, a reinforced coast line, and surprisingly green cities.

In the largest Dutch daily newspaper De Volkskrant, columnist Bert Wagendorp summarized the Panorama in one word: optimism. He mentioned the contrast with "apocalyptic prospects in which half of the Netherlands is under water and a civil war is raging the still-dry regions." Wagendorp called it "a breath of fresh air", and hopes politicians will "listen to their advice". At the same time he was sceptical, because "the advancing populism – on the left and right wing – has at least one thing in common: an unmerciful pessimism."

Sober and efficient
Alkemade is troubled by this negative sentiment in another major project: the renovation of the Binnenhof in the Hague, where the Dutch House of Representatives is located. The historical complex is not fireproof, there is a lot of overdue maintenance and all installations must be renewed. Alkemade sees the "technical" renovation – costs aimed at 475 million euro – as an opportunity to also improve the complex spatially. On his advice, 12 architects were appointed for the renovation of the various building sections, under the supervision of the coordinating architects Liesbeth van der Pol (DOK) and Ellen van Loon (OMA).

The design process – due to security reasons appointed State Secret – so far has been problematic. There was a leak to the press about the plans, and parliament members called the preliminary design "megalomaniacs". It led to DOK and OMA being removed from the project, to Alkemade's frustration. "It was not the content but the image that determined the course of events. If a space on an artist's impression looks different from the existing situation, this suggests to some that too much tax money is being spent. Visible change is read as a negative factor," he says.

"Image also plays a part literally", Alkemade further explains. "It turns out to be difficult for the members of parliament to read drawings, and thus estimate the value of the plans. But it also has to do with cultures. In France and Belgium, where I have worked, it is more obvious that as an architect you will be received with respect and trust. The mayor will immediately stand behind such a project. Nowadays, prestige is a forbidden word in the Netherlands; 'sober and efficient' is the new mantra. Which in the case of The Binnenhof, I don't think is a bad starting point."

Change
At the time of writing, the Netherlands is struggling with a "nitrogen crisis". The European standard for nitrogen emissions around nature reserves is exceeded in numerous places. At the behest of the Council of State, thousands of construction projects, including the Binnenhof renovation, have been halted. To reduce emissions, the government wants to reduce the number of livestock farms, which is met with fierce resistance from farmers. That is why it is also being examined whether it is possible to adjust the nitrogen rules. "A legal trick is being devised to solve it," Alkemade observes, "while you know that such an operation will come back as a boomerang. There is fear of change."

Change, it's a recurring theme during the conversation. "As an architect you are a specialist in change," says Alkemade, "At OMA, that was clearly the approach."

Why is change so necessary right now? To

pp. 248–249: Diagrams exhibited in Panorama Netherlands – a future perspective for the spatial planning of the Netherlands. Opposite: Photo of the traveling exhibition, Panorama Netherlands.

第248-249页："全景荷兰"：展出的图纸——荷兰空间规划的未来展望。对页："全景荷兰"巡回展照片。

比利时，建筑师获得的尊重和信任就更加明显，市长也会坚定地支持类似项目。如今在荷兰，'高级'是个禁词，'刚健质朴'成了新的口号，国会议事堂正是如此。但在这个项目里，我不认为从形象出发有多么不妥。"

改变

撰写本文时，荷兰正在"氮素危机"中挣扎。由于许多自然保护区周围的氮素排放超出了欧洲标准，应国会的要求，数以千计的工程项目停工，翻新国会议事堂也不例外。为了减少排放，政府还希望关闭一些畜牧场，但这遭到了养殖户的强烈抗议。这也是为什么荷兰现在开始考量，是否有可能调整氮素排放条例。"人们试图利用法律疏漏解决问题"，阿尔克马德观察发现，"但同时也知道这种做法需要自食其果，于是又害怕改变"。

改变，是这次采访中反复出现的一个主题。"建筑师是擅长改变的专家"，阿尔克马德说道："改变也是OMA明确的宗旨。"

为什么改变在当下如此必要？对此，阿尔克马德指出，"我们的系统已经失去了恢复力，整个国家不堪重负，生物多样性在减少，气候也发生着变化。而关键在于，人们不喜欢改变，宁可无动于衷。但如今我们正在引火烧身，必须重新看待政治制度和商业利益，将二者的位置调转。其中重要的一环，便是想象力。然而现在的问题是：我们能否用某种方式提出'改变'，让人们不仅不畏惧，还能看到它的好处，从而心生向往？"

阿尔克马德相信，"全景荷兰"能够作出回答："它联结起了所有问题。通常情况下，政治和商业各自为营，但这样一来，解决一个问题的同时，其他地方又会出现新的问题。与其抑制这种复杂性，不如串联各方利益（尽管有时彼此冲突）。建筑师就经常不得不处理类似的状况——设计理念与预算、消防要求、分区规划之间的矛盾不正是如此？这是设计工作的一部分，我们没有理由做不到。"

流程

除了想象力，同样重要的还有流程，也就是设想在怎样的情况下进行改变。阿尔克马德曾多次撰文和演讲，他撰写的《荷兰的未来：改变方向的艺术》一书，已于2020年出版。"面对全球问题时，荷兰处在了一个特殊的位置上"，他强调，"多少国家能够严肃对待气候变化？特朗普颁布的政策与气候变化背道而驰。欧盟必须负起责任，尤其是斯堪的纳维亚半岛上财富充裕、国土广阔的北欧各国，他们更有机会做点什么。"

"荷兰的特殊之处在于，它属于欧盟，理应承担责任，但人口相当密集，以至于难以找到解决方法。不过，正因如此，我们在空间规划方面有着良好的传统，常常与地方水务部门进行合作。"他经常被告知荷兰的渺小，我们的贡献不过沧海一粟，"但这不是在说荷兰的碳排放份额，而是指出一个事实：即使是荷兰这样条件不利的低洼型国家，也可以处理好气候问题。我们能够给世界其他国家提供范本，证明可行。"

下一个超级将是"通用"

阿尔克马德将在2020年下半年结束政府建筑师的任期。他打算如何收尾呢？

that, Alkemade remarks, "The resilience has disappeared from our systems because we have overloaded our country, which is now showing itself in declining biodiversity and climate change. But the thing is: people don't like change. They tend to do nothing. But now we are overplaying our own hands; political systems and business interests must be turned upside down. I see imagination as a crucial link in this. The question is: can we bring about change as an idea that evokes so much desire that not the threat but the profit comes into the picture?"

He believes Panorama Netherlands answers this idea – "It connects all questions to each other, whereas normally political and business systems are organized per sector. Then if you solve one thing, you create another problem elsewhere. Instead of suppressing the complexity we try to cross-link – sometimes conflicting – interests. Architects often have to deal with this, think of conflicts with budgets, fire department requirements and zoning plans. That is what the design profession entails; we can do that."

Storyline

In addition to imagination, it is important to develop a storyline about the context of the desired change. Alkemade wrote countless essays and speeches, and his book, *The Future of the Netherlands – a cultural – historic exploration* has been published in 2020. "The Netherlands has a unique position in relation to global questions," he states, "Because how many countries take climate change issues seriously? Donald Trump's policy goes against it. The EU does see the need to take responsibility, especially the Scandinavian countries have opportunities, as they have money and space."

"The uniqueness of the Netherlands is that it is in the EU, takes responsibility, but is very dense populated. That makes it difficult to find solutions. But what we have developed – precisely because of this – is a well-organized, rich tradition in the field of spatial planning with regional water authorities working together." He is often told that we're so small and our contribution is like a drop in the ocean. "But it's not about the share in CO_2 emissions; it's about the fact that you can also tackle climate in a difficult, low-lying country. We can offer a model for the rest of the world: that it's possible."

The next super will be the generic

His term as Government Architect ends in the second half of 2020. How does he want to end his work?

Alkemade: "I want to show how you can translate the vision for the future from Panorama Netherlands into the here and now. Under the name Panorama Local we linked seven cities to the program last November. We bring together the themes of the previous competitions – housing, care, rural areas – to implement the plans, we forge coalitions between municipalities, corporations, financiers and developers. The book, *The Future of the Netherlands* is also a form of completion. It's no major apotheosis, but an attempt to show that things have been set in motion. What I have added to discourse, is a broader way of thinking: how do you link social questions to the design profession? After the attention for Super, for extravagance, it should now be about 'the generic'."

What does he hope will retain of his term?

Alkemade: "Well, I guess the optimism, amid all the confusion and desperation that I sometimes experience. There is so much importance to do something now, to use design power. Architecture is really about something."

"我想用一种可见的形式,在此时此地表现'全景荷兰'所憧憬的未来。去年11月,我们以'全景本土'为名联结了7座城市,利用过去的竞赛主题,包括'住房''医疗''乡村发展',落实规划,并建立了市政、企业、金融家、开发商之间的联盟。《荷兰的未来》一书也是形式之一,但它不是描绘宏图大愿,而是试图说明事情开始有了进展。我所提供的,是一种更开阔的思维方式,借此思考如何将社会议题与设计实践联系起来。在'超级荷兰''奢侈'之后,是时候聚焦'通用'了",阿尔克马德坦言。

他希望在任期内留下什么?

阿尔克马德说:"我想,是乐观。有时即使充满困惑和沮丧,我仍会保持乐观。在当下,运用设计的力量,着手做一些事,极其重要。而建筑,不容小觑。"

Kirsten Hannema profile, refer to p. 19.

Architect **Floris Alkemade** is one of the leading voices in the current architecture debate. After graduating from TU Delft – with Rem Koolhaas as his mentor - he worked at OMA for 18 years, after which he started his own office, FAA. In September 2015 he was appointed Chief Government Architect, and in this role became the face and conscience of Dutch architecture. He organized, amongst others, Open Calls around societal issues such as (the shortage of) affordable housing. In addition, as of 2014, he leads the Tabula Scripta electorate at the Amsterdam Academy of Architecture and was co-curator of the International Architecture Biennale Rotterdam 2018, The Missing Link, on the "essential, yet missing link that should enable us to make the necessary transition to a climate-proof way of life". He was elected Architect of the Year 2018 by the professional community. His book *The Future of the Netherlands – the art of changing direction*, has published in 2020.

克丝汀·汉内马的个人资料,参见第19页。

建筑师弗洛里斯·阿尔克马德是目前建筑界的领袖之一。他曾就读于代尔夫特理工大学,师从雷姆·库哈斯,毕业后在OMA工作了18年。之后他成立了自己的工作室FAA。2015年9月,他被任命为首席政府建筑师,代表着荷兰建筑的门面和良心。任职期间,他曾为解决经济型住房短缺等社会问题组织设计竞赛。除此之外,2014年他牵头了阿姆斯特丹建筑学院的"Tabula Scripta"项目,2018年作为鹿特丹国际建筑双年展的联合策展人,策划了展览"缺失的联系"(The Missing Link),关注"转型生活方式以应对气候变化的过程中丢失的重要环节"。他还被同行推选为2018年度建筑师。他撰写的《荷兰的未来:改变方向的艺术》已于2020年出版。

建筑师简介

Peter van Assche transitioned into architecture after a succesful career in experimental mathematics and is now the founding principal of bureau SLA, an Amsterdam-based firm focused on the necessity of transitioning to a circular economy through design. As an office, **bureau SLA** consists of a team of architects and builders, supported by architectural historians, landscape architects, and energy experts. The studio does not wait for commissions to be given, but builds and develops in the city in an innovative way – from their own initiatives and with their own manpower. With this trial and error mentality, Peter van Assche discovers the full potential of material use, energy, waste flows, smart living and working, and development processes.

彼得·范·阿什在应用数学方面取得成功后转入建筑业,现在是**布洛 SLA**的创始兼负责人。该事务所总部位于阿姆斯特丹,专注于探索通过设计向循环经济过渡的必要性。作为事务所,**布洛 SLA**团队由建筑师和建造商共同组成,并由建筑历史学家、景观建筑师和能源专家提供支持。工作室从不等着工作自己找上门,而是靠主动性和人力,以一种创新的方式在城市中寻求建设和发展。通过这种反复试验的思维方式彼得·范·阿什充分发掘了材料使用、能源、废物流通、智能生活和工作、以及开发过程的潜力。

Architect Mathijs Cremers is **ZakenMaker**. Having spent 6 years gaining experience as a professional architect, he set up his own business ZakenMaker. Here, he finds the freedom and flexibility needed to combine knowledge of designing with knowledge of manufacturing: a case of "know why meets know-how". ZakenMaker designs, realises and experiments with projects commissioned by himself or others.

建筑师马蒂亚斯·克雷默斯即是**扎克马克**。他花了6年时间积累专业建筑师的经验,而后创立了自己的公司扎克马克。在这里,他找到了将设计知识与制造知识相结合所需的自由度和灵活性,即"知为何且知何为"的案例。扎克马克设计、实现和试验自己或他人委托的项目。

Space&Matter is a multidisciplinary spatial design practice based in Amsterdam. They merge architecture, urban planning, technological platforms and systems thinking to constantly push the envelope of the spatial discipline. Space&Matter specializes in unconventional projects which connect people with each other, and their built environment. Underlying their agenda is a focus on community, sustainable foodscapes, and circularity. From rooftop farms to floating residential developments, Space&Matter works on a small scale to explore innovative solutions to global challenges. They experiment with renewable energies, decentralized, and off-grid technologies to close resource loops and create circular and resilient living environments.

空间&变体建筑事务所是一个跨界空间设计事务所,总部设在阿姆斯特丹。他们融合了建筑设计、城市规划、技术平台和系统思维,以不断突破空间学科的界限。事务所专长于非传统项目,这些项目将人与人、人与建筑环境联系在一起。他们对社区、可持续的食物景观和循环性的重视潜藏在他们的日常工作中。从屋顶农场到漂浮住宅开发,空间&变体建筑事务所以小尺度开展工作,旨在探索出创新的解决方案来应对全球性挑战。他们对可再生能源、分散和离网技术进行试验,形成资源闭环,并创造可循环且能迅速恢复的生活环境。

Hilberinkbosch Architects, founded in 1996 by Annemariken Hilberink (photo) and Geert Bosch, is passionately looking for the smartest solutions for complex tasks. Their projects range from private and project-based housing, offices, renovations to large-scale studies. The firm attaches great importance to having dialogues with all parties involved to achieve optimal results and sees consultation as an important part of the process. Through the adaptation of local qualities, their work evokes a sense of recognition, but at the same time, the new expression and unexpected use of material seem to alienate. Their preference for a big gesture and poetry of life creates robust architecture with sensory experiences.

希尔伯林克·博世建筑师事务所由安娜玛丽肯·希尔伯林克（左图）和格尔特·博世于1996年创立，一直在热切地寻求针对复杂任务最智能的解决方案。他们的项目从私人住宅、政府资助的集合住宅、办公室、翻新项目到大规模研究，跨域广泛。事务所十分重视与项目相关的各方进行对话，以达成最优方案，并将讨论视为设计过程的重要组成部分。他们的作品通过对当地文化的转译，唤起一种认同感，但与此同时，新的表达方式和意想不到的材料运用又彼此独立。他们对主体形态和诗意栖居的偏爱创造了感官体验丰富的有力量的建筑。

RAU Architects is a leader in the realization of circular construction projects. RAU looks to push boundaries with the help of the latest technologies and from new perspectives. Everything is temporary, but the consequences of impermanence are permanent. Cyclical thinking is self-evident for RAU. It defines a fundamentally different approach to a housing challenge: not resulting in a self-contained building but an active element in a circular system. At the level of the area, the energy, the material, society, culture, nature, realization and the future: boundaries are blurred, and synergy is created. Thomas Rau ranks top on the "Top 100 most sustainable companies" (De Duurzame 100).

RAU 建筑师事务所是实现循环建设项目的领导者。RAU希望借助最新的技术和全新的视角来突破设计界限。一切都是暂时的，但暂时性的影响却是永恒不变的。周期性思维对于RAU来说是不言而喻的。它定义了一种从根本上就与众不同的、应对住房问题的方法；他们创造的不是一个自给自足的建筑，而是一个激活所有元素的循环系统。在区域层面上，能源、物质、社会、文化、自然、认知和未来的界限都将模糊，并产生协同效应。托马斯·劳的事务所在"最具可持续性的100家公司"排名中占据首位。

RO&AD Architects is an architectural office located in the Netherlands. They are acting not only on the scale of buildings but also on the scale of systems where the buildings sit in. Nature, too, works that way. Every species is connected and dependent on dozens or even hundreds of others. Ideally, they create buildings that would positively act in these ecosystems to create bigger, positive footprints, and to add values like happiness and fun.

RO&AD 建筑师事务所是一家设在荷兰的建筑事务所。他们不仅在建筑尺度上下功夫，还在建筑所处的系统尺度上下功夫。自然界亦是如此。每个物种都是相互联系的，并依赖于数十个甚至数百个其他物种。理想情况下，他们创造的建筑将在这些生态系统中发挥积极作用，并扩大积极影响，以增加诸如幸福、快乐此类的价值。

255

RAAAF operates at the crossroads of visual art, architecture, and philosophy. The studio started in 2006 by partners: Prix de Rome laureate Ronald Rietveld and philosopher (right) Erik Rietveld (left). Through a unique working method based on multidisciplinary research with scientists and craftsmen, their real-life thinking models link local qualities with the past, present, and future.

Atelier de Lyon was started by artist Erick de Lyon (no photo). His work finds its expression in the Dutch landscape. The starting point of the work is obtaining knowledge about the context. Out of this knowledge solutions often emerge "almost spontaneously". Knowledge is acquired by combining observations of the environment with relevant knowledge of others.

RAAAF 处在视觉艺术、建筑与哲学的交叉领域。该事务所于 2006 年由合伙人——罗马大奖（Prix de Rome）得主罗纳德·里特维尔德（右）和哲学家埃里克·里特维尔德（左）创立。通过与科学家和工匠合作进行多学科研究的独特工作方式，他们利用现实思维模式将地方特性与过去、现在和未来关联起来。

里昂工作室由艺术家埃里克·德·里昂创立（无照片）。他的作品从荷兰的风景中汲取情感。项目的出发点是从环境中观察获取信息。在这些信息中，解决方案往往是"几乎自发地"显现出来。这些信息则通过将对环境的观察和从第三方获取的相关信息结合而获得的。

Robert A.M. Stern Architects, LLP, is a 265-person firm of architects, interior designers, and supporting staff. Over its 50-year history, the firm has established an international reputation as a leading design firm with wide experience in residential, commercial, and institutional work. As the firm's practice diversifies, its geographical scope widens to include projects in Europe, Asia, South America, and throughout the United States. The firm maintains an attention to detail and commitment to design quality which has earned international recognition, numerous awards, and citations for design excellence, including National Honor Awards of the American Institute of Architects, and a lengthening list of repeat clients.

罗伯·A.M. 斯坦恩建筑师事务所是一家由建筑师、室内设计师和辅助人员组成的事务所，共有员工 265 人。历经 50 年，该事务所已成为国际领先的设计事务所，在住宅、商业和公共机构领域都拥有丰富的经验。随着事务所业务的多元化，其业务的地域范围也扩大到欧洲、亚洲、南美洲以及北美洲的整个美国。该事务所一直保持着对细节的重视和对设计品质的承诺，这也使它获得了国际认可，揽获众多奖项和卓越设计表彰，其中包括美国建筑师协会的国家荣誉奖，同时也获得不断增加的老客户。

Shortly after finishing his architecture studies at the Vrije Universiteit Brussel (Belgium), Frederik Vermeesch started working as an architect at **Rijnboutt** based in Amsterdam. Since 2009 he is one of the owners of the firm, which practice involves architecture, urbanism and landscape.

弗雷德里克·维米希在比利时布鲁塞尔自由大学完成建筑学业后不久，就开始在阿姆斯特丹的**里恩博特建筑师事务所**担任建筑师。自 2009 年起，他成为该公司的所有者之一。该公司的业务涉及建筑设计、城市规划以及景观设计。

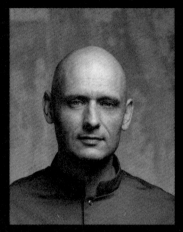

Floris Schiferli is a partner and principal architect at **Superuse Studios** since 2013. He is working on new processes within architecture and urban planning based on urban metabolism with a focus on concept and realization, always searching for the challenges in legislation, regulations and cooperation processes. **Superuse Studios**, a pioneer in circular building, designs building with residual flows to create blue, circular, sustainable architecture and strategies. They have developed a series of tools to design and build with residual flows, with the aim of impact. Since 2020, Floris is also a member of the board of Oogstkaart, setting up new network possibilities for circular building.

弗洛里斯·谢弗里自2013年起担任**超级利用工作室**的合伙人兼首席建筑师。他关注建筑和城市规划中基于城市新陈代谢的新流程，专注于概念和实现，并一直在法律、法规和合作过程中寻求挑战。**超级利用工作室**是循环型建筑领域的先锋，设计出了具有"剩余循环"可能的建筑，以创造有水循环、环境循环、可持续的建筑和策略。他们开发了一系列工具用以设计和建造具有"剩余循环"可能的建筑，以获得影响力。自2020年起，弗洛里斯成为奥格斯特卡特（一个收集了各种废料数据，旨在对其进行再利用的网站）董事会成员之一，为循环建筑建立新的网络可能性。

Neutelings Riedijk Architects was founded in 1989 by Willem Jan Neutelings (photo) and Michiel Riedijk, complemented in 2006 with managing director Carl Meeusen. The office is based in Rotterdam and offers a strong commitment to design excellence, realizing high-quality architecture by developing powerful and innovative concepts into clear form. The office has received awards such as the BNA Kubus, the Belgian Building Award, and the Rotterdam Maaskant Prize and has been shortlisted for the Mies van der Rohe Award. Michiel Riedijk and Willem Jan Neutelings are recognized as International Fellows of the RIBA.

诺特林斯·里迪克建筑师事务所由威廉姆·扬·诺特林斯（照片）和米歇尔·里迪克于1989年创立，其常务董事卡尔·梅森于2006年加入。办公地点设在鹿特丹，致力于精益求精的设计，通过将强有力且创新的理念发展为具象形式，来实现高品质的建筑。该事务所曾获"BNA Kubus"（荷兰皇家建筑师学会授予的最高荣誉）、"比利时建筑奖"，以及"鹿特丹马斯坎特奖"等奖项，并入围了"密斯·凡·德·罗奖"。米歇尔·里迪克和威廉姆·扬·诺特林斯是英国皇家建筑师协会（RIBA）的国际会员。

CIVIC Architects is an office for public architecture. They design libraries, bridges, cultural buildings, public forums, squares, research and education buildings, housing, sculptures, and stations. In the dialogue of social issues and intra-architectural matter (form, structure, tectonics) resides the true relevance of architecture. Good architecture solves problems and exceptional architecture combines this with the sublime. They believe in history as a continuously changing flow and in connecting scales, ranging from urbanism to detailing. For them, architecture is a way to turn theoretical ambitions into physical reality. CIVIC Architects strives for powerful, generous and location-specific architecture that is built to stand the test of time.

城市建筑事务所是一家公共建筑事务所。他们设计范围广泛，包括图书馆、桥梁、文化建筑、公共集会场所、广场、教育研究建筑、住宅、雕塑和车站等。他们认为在社会问题与建筑内部问题（形态、结构、构造）的对话中，存在着建筑的真正意义。好的建筑可以解决问题，而卓越的建筑则将解决问题与崇高精神相结合。他们相信历史会被不断地创造，并相信从城市化到细节的各个尺度之间的关联性。对他们来说，建筑是一种将理论抱负变为现实的方式。城市建筑事务所致力于创造强大、慷慨、具有地域性且经得住时间考验的建筑。

After many years of experience in cultural, housing and urban projects, Patrick Koschuch now runs his own practice. **Koschuch Architects** strive for a perfectly working, rational building with unambiguous organization, clear floor plans, and optimum building exploitation. This architectural rational exuberance goes beyond solving specific needs and making unique iconic gestures. The aim is to realize buildings that offer added value for both users and the environment. Their approach see architecture as an iterative design process that leads to a strong concept, where people can embrace their building.

经过多年在文化、住宅和城市项目领域积累的经验，帕特里克·科舒奇如今经营着自己的事务所。**科舒奇建筑师事务所**致力于通过明确的组织方式，清晰的平面图和建筑的最优化利用来实现完美运作的理性建筑。这种建筑上的理性的充沛不仅限于满足特定需求和做出独特的标志性形态，他们旨在实现能为用户和环境提供附加价值的建筑。他们将建筑设计的方法视为一个迭代的设计过程，从中产生了一个强大的理念，在他们的设计中，人们可以拥抱建筑。

KAAN Architects is a Rotterdam, São Paulo and Paris based architectural firm that operates in a global context while merging practical and academic expertise within the fields of architecture, urbanism, and research on the built environment. Led by Kees Kaan (center), Vincent Panhuysen (right) and Dikkie Scipio (left), the studio consists of an international team of architects, landscape architects, urban planners, engineers, and graphic designers. KAAN Architects believes in cross-pollination between projects and disciplines as an essential tool to fostering a critical debate within the firm. The firm expanded its international presence with satellite offices in São Paulo (2015) and Paris (2019), led by Renata Gilio and Marylène Gallon respectively.

KAAN 建筑师事务所是一家在全球范围内运营的建筑事务所，在鹿特丹、圣保罗和巴黎都设有分部，事务所融合了建筑、城市规划及建筑环境研究领域的实践和学术专长。该事务所由卡恩·基斯（中）、文森特·潘惠森（右）和迪基·西皮奥（左）领导，是一支由建筑师、景观设计师、城市规划师、工程师及平面设计师组成的国际化团队。KAAN 建筑师事务所将项目与学科之间的互促互益视作在公司内部引发批判性辩论的重要手段。事务所通过由雷纳塔·吉利奥和马里兰·加仑分别领导的圣保罗（2015）和巴黎（2019）分部扩大了它的国际知名度。

Frits van Dongen assembled a portfolio of cultural projects, including theatres and concert buildings, and delivering over 12,000 housing units. Each project acts as a specific response to its brief and its physical and social context. Through an extensive and multi-disciplinary network, **Frits van Dongen Architects and Planners** is able to design and realize projects of significant complexity.

弗里茨·范·东恩建筑规划事务所完成了一系列包括剧场、音乐厅在内的文化类项目，并交付了超过 1,2000 套住宅。每个项目都针对其任务书及物理和社会性文脉作出了明确的回应。基于广泛且多学科的网状系统，事务所的建筑师和规划师们总能设计并实现复杂性极高的项目。

NL Architects is an Amsterdam based office. Their 3 principals, Pieter Bannenberg (left), Walter van Dijk (center) and Kamiel Klaasse (right) graduated from the Technical University in Delft. The trio officially opened their practice in 1997. NL Architects aspires to catalyze urban life. The office is on a constant hunt to find alternatives for the way we live and work. They understand architecture as the speculative process of investigating, revealing and reconfiguring the wonderful complexities of the world we live in. Their projects focus on ordinary aspects of everyday life, including the unappreciated, to bring to the fore the unexpected potential of things surrounding us.

NL建筑师事务所总部位于阿姆斯特丹。他们的三位负责人,彼得·班纳伯格(左)、沃尔特·范·迪克(中)和卡米尔·克拉斯(右)都毕业于代尔夫特理工大学。三人于1997年正式成立了自己的事务所。NL建筑师事务所渴望催化城市生活,一直在寻求我们生活和工作的另一种方式。他们把建筑理解为研究、揭示和重新配置我们所生活的、奇妙复杂的世界的思辨过程。他们的项目着眼于平凡的日常生活,亦不忽视那些未能为人欣赏的,从而将我们周围事物所蕴含的出人意料的潜能展现在世人面前。

XVW Architecture was founded in 2010 by Xander Vermeulen Windsant. Before that, he studied at the Faculty of Architecture, University of Technology Delft and worked at Clausen Kaan Architects in Amsterdam. The small team, based in Amsterdam, focuses mostly on residential projects, ranging from small private commissions, medium size newly built apartment buildings to large scale renovation projects. Context, Transformation, and Collaboration, with its inhabitants, drive the design work at the studio. With its hands-on approach, the studio is asked by private clients and commercial developers alike to work on projects in which these themes define the character of the project.

XVW建筑师事务所由桑德·维尔穆林于2010年创立。在此之前,他就读于代尔夫特理工大学建筑学院,并在阿姆斯特丹的克劳斯·卡恩建筑师事务所工作。这个位于阿姆斯特丹的小团队主要专注于住宅项目,尺度从小型私人委托、中型公寓新建到大型改造项目不等。文脉、改善以及与当地居民的协作推动着事务所的设计工作。基于对事务亲身实践的设计方法,他们会受邀设计一些由私人客户或商业开发商委托的、通过上述主题来明确其特征的项目。

Happel Cornelisse Verhoeven handles a broad portfolio of public buildings, redevelopments, residential buildings, and public interiors. Given the firm's marked interest in the cultural-historical layering of cities and buildings, most of its projects are located at the interface between old and new. The practice's approach is one of regenerative synthesis, juxtaposing past and present in such a way that they reinforce one another's special qualities. The partners, Ninke Happel (right), Floris Cornelisse (center) and Paul Verhoeven (left) lead the practice based on their different but complementary backgrounds. HCVA recently founded Het Rotterdams Woongenootschap, a cooperative association that initiates, manages and leases housing projects in the central districts of Rotterdam. Photo by Willem De Kam.

哈佩尔-科尼利斯-韦尔霍芬建筑师事务所完成过各类公共建筑、重建、住宅和公共室内设计项目。因该事务所对城市与建筑的文化历史表现出浓厚的兴趣,其大部分项目都位于新与旧的交界处。该事务所注重综合再生的建造方式,即以某种方式将过去与现在并置,以增强彼此的独特品质。合伙人尼克·哈佩尔(右),弗洛里斯·科尼利斯(中)和保罗·韦尔霍芬(左)以各自不同但互补的背景来领导事务所。事务所最近成立了鹿特丹房屋公司,这是一个在鹿特丹市中心地区发起的、管理和租赁住房项目的合作协会。

Joost Ector graduated with honors from the Eindhoven University of Technology in 1996 and immediately afterward joined the Rotterdam architect Jan Hoogstad. In 2002, he became co-owner of the company which shortly thereafter changed its name to **Ector Hoogstad Architects**. The agency currently has 5 managing partners with around 50 employees and designs for almost all sectors, with an emphasis on complex, often multifunctional public design assignments. In recent years, work has been done on large-scale projects for various Dutch universities, cultural buildings, a large number of transformation projects and the "biggest bicycle parking facility in the world".

乔斯特·埃克特于 1996 年以优异的成绩毕业于埃因霍温理工大学，随后立即加入了鹿特丹建筑师贾恩·胡格斯塔德的团队。2002 年，他成为事务所的合伙人，此后不久公司更名为**埃克特·胡格斯塔德建筑师事务所**。该机构目前由 5 位合伙人管理，约有 50 名员工，设计几乎涵盖了所有领域。其项目强调复杂性，常被委托多功能的公共设计任务。近年来，他们完成了荷兰多所大学的大型项目、文化建筑、大量的改造项目，以及"世界上最大的自行车停放设施"。

Francine Houben formulated 3 fundamental concepts of her architectural vision during her studies at the Delft University of Technology. Designing primarily for People, constructing spaces relevant to Place, and forging connections that give a building Purpose, have remained consistent. They are the underlying values of **Mecanoo**'s practice over the past 3 decades. Always seeking inspiration in the details of specific sites and locations. Francine bases her work on precise analysis. She interweaves social, technical, playful and human aspects of space-making to create a unique solution to each architectural challenge.

在代尔夫特理工大学学习期间，法兰馨·侯班提出了三个建筑构想的基本概念：以人为本做设计，构建与场所相关的空间，以及为建筑目的建立联系，并始终践行。这也是**麦肯诺建筑师事务所**在过去的 30 年里一直秉持的基本价值观。法兰馨总是在特定基址和地点的细节中寻求灵感，她的工作基于精确的分析。她将社会性、技术性、趣味性和人性化的空间营造交织在一起，为每次挑战创造出其独特的解决方案。

Koen van Velsen Architects consists of a multidisciplinary team working together on projects ranging from architecture to urban design. Koen van Velsen received numerous distinctions for his work, such as Mart Stam Award from the city of Amsterdam, Rietveld Award from the city of Utrecht and the BNA cube, Oeuvre award from the Association of Dutch Architect. Between 2009 and 2015, Koen van Velsen was Dutch Railroad Architect on behalf of NS and ProRail. He strived to create better cohesion between train stations and the surrounding public realm. Koen van Velsen was acting Government Architect and part of the Committee of Government Advisors between January to September 2015.

科恩·凡·威尔森建筑师事务所由一支多学科的团队组成，致力于从建筑设计到城市设计的各个项目。科恩·凡·威尔森的作品荣获了许多奖项，例如，阿姆斯特丹市的马特·斯坦奖、乌得勒支市的里特菲尔德奖和荷兰建筑师协会颁发的 BNA 立方奖。2009 年至 2015 年间，科恩·凡·威尔森代表 NS 和 ProRail 担任荷兰铁路建筑师。他努力在火车站和周围公共空间之间建立更好的凝聚性。2015 年 1 月至 9 月间，科恩·凡·威尔森曾担任代理政府建筑师和政府顾问委员会成员。

Antonio Cruz (left) and **Antonio Ortiz** (right) graduated from the School of Architecture of Madrid in 1971 and have since worked together. They taught at the EPFL Lausanne, ETH Zurich, Harvard University, Cornell University, Columbia University, and the University of Navarra. They have held the Kenzo Tange Professorship at Harvard's GSD, and are honorary professors at the University of Seville. They have received numerous awards, such as the Honorary Fellowship by the American Institute of Architects, Knight Grand Cross – the Order of the Netherlands Lion awarded by King William I of the Netherlands, Gold Medal of Architecture awarded by the Superior Council of Colleges of Architects of Spain.

安东尼奥·克鲁兹（左）和**安东尼奥·奥尔蒂斯**（右）于1971年毕业于马德里建筑学院，此后一直在一起工作。他们曾任教于洛桑联邦理工学院、苏黎世联邦理工学院、哈佛大学、康奈尔大学、哥伦比亚大学和纳瓦拉大学。他们曾在哈佛大学设计研究生院担任客座教授，并在塞维利亚大学担任名誉教授。他们获得了无数奖项，例如美国建筑师协会荣誉院士、荷兰国王威廉一世授予的荷兰狮子大十字勋章、西班牙建筑师学院高级理事会授予的建筑金奖等。

380+ 建筑作品　　5+家具艺术品　　1500+摄影作品
500+ 手绘作品 100+ 论文及安藤故事 10+ 展览及建筑小品

致敬20

基本信息:开本 16开 / 尺寸 215mm×280mm / 语言 內文全中文, 索引中日英三语 / 页码 3496页(以最终印刷版为准) / 结
国际建筑联盟 / 主编 马卫东/执行编译 安藤忠雄全集编辑部/出版发行 中国建筑工业出版社/主要作(译)者 安藤忠雄 肯尼斯

中日邦交正常化50周年纪念项目　日本国际交流基金会赞助项目

安藤忠雄全集
TADAO ANDO COMPLETE WORKS

世纪传奇建筑家　全解建筑世界里的光影挑战

四色全彩 / 装帧 精装 / 发行范围 中国 / 内容监修 安藤忠雄建筑研究所 / 特别支持 日本新建筑社 / 书籍策划 文筑国际 IAM
铃木博之　彼得·艾森曼　松叶一清　中川 武　三宅理一　朱涛　马卫东等/本书相关消息敬请关注官方微信公众号"安藤忠雄之家"

©安藤忠雄建筑研究所

Spotlight:
Aurora Museum
Tadao Ando Architect & Associates

特别收录：
震旦博物馆
安藤忠雄建筑研究所

The AURORA building is located on the east bank of the Huangpu River in Shanghai, inside the built Aurora Plaza in the Pudong area, which is full of skyscrapers. We needed to transform the lower part of the building to give it a new mission as a museum.

The upper part of the Aurora Plaza is wrapped in a glittering exterior that transforms into a giant screen at night, a symbol of the dynamism of the ever-changing city of Shanghai.

In contrast to this vivid and dynamic effect, we set the new museum as a quiet light box with wonderful internal light = "gem box". It is located in the most prosperous scenery of Pudong area, where various buildings of different styles are lined up. The low-key expression of the glass curtain wall with the whole LED lighting embedded in it should make a different presence here.

The interior design also presents the space of "gem box", where all the precious exhibits are packed in the most visually beautiful form. The first thing that came to our mind was a glass case that made each exhibit appear to be floating in mid-air. All the details are made extremely infinitesimal, and the proportion and layout of exhibits are determined by the structural columns of the building and the span in the middle. We would like to use this transparent geometry to create a space for exhibition. Regarding the presentation of the exhibition space of the museum.

The biggest problem encountered in this design is that the original building was built according to the standard of an office building. The space structure is average, and the ceiling height required by the gallery space cannot be achieved. Based on this problem, we believe that the space should not be broken by adding partitions. Instead, the exhibition space should be regarded as a whole, in which several characteristic spaces enclosed by exhibition hall and buffer space could be inserted. The spatial form of the box structure allows the restricted space to reach its maximum possible performance. The result come out are the newly inserted cantilevered staircase depicting a 1/4 arc rising vertically, the exhibition hall with cantilevered ceiling connecting the third and fourth floors, and the platform on the second floor covered with greenery.

The new Aurora Museum A2 space on the second and third floors of the high-rise buildingopened In 2021.

In contrast to the original Aurora Museum, where the elegant Gem Box and historical items are permanently exhibited, the new Aurora Museum A2 space is flexible and a temporary display area for contemporary works of visual art, design, fashion, and performance, with the gorgeous two-story exhibition space designed with walls of gentle curvature. We look forward to the excellent and unique Gem Box and AURORA space becoming the new landmark in the center with great potential of Shanghai.

Text by Tadao Ando Architect & Associates

Credits and Data
Project title: Aurora Museum
Location: Shanghai, China
Design: 2009.09-2012.02 (A1)
    2018.05-2021.09 (A2)
Construction: 2010.11-2012.02 (A1)
     2019.09-2021.09 (A2)
Structure: Reinforced Concrete
Fuction: Museum of art
Site area: 9,777.00 m^2
Total floor area: 6,316.00 m^2 (A1)
      3,461.00 m^2 (A2)

pp. 264–265: View of the Shanghai skyline from the entrance of the museum. Opposite: See the exterior of the building from the passage that is continuous with the entrance. The building is designed to look like a "gem box", a box filled with quiet light.

第 264-265 页：从博物馆入口眺望上海天际线景观。对页：从与入口连续的通道看向建筑外观。建筑被设计成"宝石箱"的样子，一个充满静谧光线的盒子。

基地位于上海黄浦江东岸，已建成的震旦大楼矗立在超高层建筑林立的浦东新区。项目需要将大楼的低层部分加以改造，赋予其美术馆的新使命。

震旦大楼的高层部分被金光闪闪的外部装饰所包裹，夜晚时分，整面化身为巨大的屏幕，成为日新月异的上海这座城市的活力的象征。

与这种生动又富有活力的效果相映衬，我们将低层的美术馆考虑设定为一个由内而外散发美妙光线的、静谧的光之盒——"宝石盒"。它位于浦东最繁华的风景之中，那里高耸着林林总总不同样式的高楼大厦，这种整面嵌入LED照明后所形成的玻璃幕墙外观，低调克制的表现形式，或许会在这里表现出了不一样的存在感。

室内的设计概念同样是创造一个能将种种珍贵展品以视觉上最美的形式包裹起来的"宝石盒"空间。由此，我们首先想到的是，让每一个展品看上去都仿佛悬浮在空中一样的玻璃展柜。所有细部都做到极小，结合建筑物的结构柱及中间的跨度来确定展品比例与布局。以这种透明的几何学来打造展示空间。

本次设计遇到的最大问题是原建筑最初是作为办公楼而建的，空间结构呈现均质化，并且无法满足美术馆空间所需要的层高。对此，我们认为不能切碎空间，而应将全部展示区看做一个空间整体，然后在曲中插入几个被展厅和缓冲空间围合起来的富有个性的空间，基于这种套匣结构的空间形式，在一个被限制的空间中实现最大化的演绎。也就是说，在垂直方向上插入的一个1/4圆弧螺旋上升的挑空楼梯，连接起三四层的曲面顶的展厅，以及二层展厅的被绿植覆盖的平台。

2021年，我们在高层栋二三层新设计的"震旦博物馆A2空间"（Aurora Museum A2 space）也开馆了。

相较原有的震旦博物馆——一个静谧的、长期展示历史文物的"宝石盒"，"A2空间"被冠以"AURORA"（震旦）之名，追求临时展示视觉艺术、设计、时尚、表演等现代作品的动态灵活空间。为了回应这一点，我们设计了一个拥有光曲面墙壁的亮晶晶的双层展览空间。

我们期待这一组位于上海市中心高潜力地块的美丽且富有个性的"宝石盒"与"AURORA"，能成为新的城市地标。

安藤忠雄建筑研究所/文

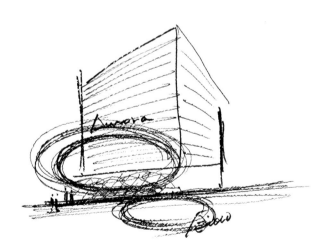

This Page: Tadao Ando manuscript. Opposite, above: Interior view of Buddhism exhibition area on the sixth floor. Opposite, below: The view of Huangpu River in front of the exhibition space.

本页：安藤忠雄手稿。对页，上：六层佛教展示区域内景；对页，下：在展陈空间远眺面前的黄浦江风景。

269

This page, Opposite: Interior view of area added in 2021. Compared with the main museum hall, which displays traditional Chinese art, the additional space can be flexibly used to display modern art and fashion shows. This impressive space of wavy walls and light reflects the image of AURORA. All images on pp. 264–271 by Aurora Museum.

本页,对页:2021年增建区域内景。相较于展示中国传统艺术的本馆,增建空间可灵活运用于展示现代艺术或作为时尚秀场等。这个让人印象深刻的波浪墙面与光的空间,映照了"AURORA"(震旦)的形象。